Wholeness and the implicate order

By the same author

Causality and Chance in Modern Physics (Routledge & Kegan Paul, 1957)
Quantum Theory (Prentice-Hall, 1951)
The Special Theory of Relativity (Benjamin, 1965)

Wholeness and the implicate order

David Bohm

Professor of Theoretical Physics
Birkbeck College
London

ROUTLEDGE & KEGAN PAUL
London, Boston and Henley

First published in 1980
by Routledge & Kegan Paul Ltd
39 Store Street, London WC1E 7DD,
9 Park Street, Boston, Mass. 02108, USA and
Broadway House, Newtown Road,
Henley-on-Thames, Oxon RG9 1EN
Set in Times
by Input Typesetting Ltd, London
and printed in Great Britain by
Lowe & Brydone Ltd
Thetford, Norfolk
Reprinted with corrections 1980

British Library Cataloguing in Publication Data

Bohm, David

Wholeness and the implicate order.
1. Quantum theory
2. Physics – Philosophy
I. Title
530.1'2 QC174.13 80-40026

ISBN 0 7100 0366 8

Contents

Acknowledgments

The author and publisher would like to thank the following for permission to reproduce copyright material: The Van Leer Jerusalem Foundation (chapters 1 and 2, from *Fragmentation and Wholeness*, 1976); the editors of *The Academy* (chapter 3, from *The Academy*, vol. 19, no. 1, February 1975); Academic Press Ltd (chapter 4, from *Quantum Theory Radiation and High Energy Physics*, part 3, edited D. R. Bates, 1962); Plenum Publishing Corporation (chapters 5 and 6, from *Foundations of Physics*, vol. 1, no. 4, 1971, pp. 359–81 and vol. 3, no. 2, 1973, pp. 139–68).

Introduction

This book is a collection of essays (see Acknowledgments) representing the development of my thinking over the past twenty years. A brief introduction will perhaps be useful in order to indicate what are the principal questions that are to be discussed, and how they are connected.

I would say that in my scientific and philosophical work, my main concern has been with understanding the nature of reality in general and of consciousness in particular as a coherent whole, which is never static or complete, but which is in an unending process of movement and unfoldment. Thus, when I look back, I see that even as a child I was fascinated by the puzzle, indeed the mystery, of what is the nature of movement. Whenever one *thinks* of anything, it seems to be apprehended either as static, or as a series of static images. Yet, in the actual experience of movement, one *senses* an unbroken, undivided process of flow, to which the series of static images in thought is related as a series of 'still' photographs might be related to the actuality of a speeding car. This question was, of course, already raised in essence philosophically more than 2,000 years ago in Zeno's paradoxes; but as yet, it cannot be said to have a satisfactory resolution.

Then there is the further question of what is the relationship of thinking to reality. As careful attention shows, thought itself is in an actual process of movement. That is to say, one can feel a sense of flow in the 'stream of consciousness' not dissimilar to the sense of flow in the movement of matter in general. May not thought itself thus be a part of reality as a whole? But then, what could it mean for one part of reality to 'know' another, and to what extent would this be possible? Does the content of thought merely give us abstract and simplified 'snapshots' of reality, or can it go further, somehow to grasp the very essence of the living movement that we sense in actual experience?

It is clear that in reflecting on and pondering the nature of movement, both in thought and in the object of thought, one comes inevitably to the question of wholeness or totality. The notion that the one who thinks (the Ego) is at least in principle completely separate from and independent of the reality that he thinks about is of course firmly embedded in our entire tradition. (This notion is clearly almost universally accepted in the West, but in the East there is a general tendency to deny it verbally and philosophically while at the same time such an approach pervades most of life and daily practice as much as it does in the West.) General experience of the sort described above, along with a great deal of modern scientific knowledge concerning the nature and function of the brain as the seat of thought, suggest very strongly that such a division cannot be maintained consistently. But this confronts us with a very difficult challenge: How are we to think coherently of a single, unbroken, flowing actuality of existence as a whole, containing both thought (consciousness) and external reality as we experience it?

Clearly, this brings us to consider our overall *world view*, which includes our general notions concerning the nature of reality, along with those concerning the total order of the universe, i.e., *cosmology*. To meet the challenge before us our notions of cosmology and of the general nature of reality must have room in them to permit a consistent account of consciousness. Vice versa, our notions of consciousness must have room in them to understand what it means for its content to be 'reality as a whole'. The two sets of notions together should then be such as to allow for an understanding of how reality and consciousness are related.

These questions are, of course, enormous and could in any case probably never be resolved ultimately and completely. Nevertheless, it has always seemed important to me that there be a continuing investigation of proposals aimed at meeting the challenge that has been pointed out here. Of course, the prevailing tendency in modern science has been against such an enterprise, being directed instead mainly toward relatively detailed and concrete theoretical predictions, which show at least some promise of eventual pragmatic application. Some explanation of why I want to go so strongly against the prevailing general current seems therefore to be called for.

Aside from what I feel to be the intrinsic interest of questions that are so fundamental and deep, I would, in this connection, call attention to the general problem of fragmentation of human consciousness, which is discussed in chapter 1. It is proposed there that the widespread and pervasive distinctions between people (race, nation, family, profession, etc., etc.), which are now preventing mankind from working together for the common good, and indeed, even for survival, have one of the key factors of their origin in a kind of thought that treats *things* as inherently divided, disconnected, and 'broken up' into yet smaller constituent parts. Each part is considered to be essentially independent and self-existent.

When man thinks of himself in this way, he will inevitably tend to defend the needs of his own 'Ego' against those of the others; or, if he identifies with a group of people of the same kind, he will defend this group in a similar way. He cannot seriously think of mankind as the basic reality, whose claims come first. Even if he does try to consider the needs of mankind he tends to regard humanity as separate from nature, and so on. What I am proposing here is that man's general way of thinking of the totality, i.e. his general world view, is crucial for overall order of the human mind itself. If he thinks of the totality as constituted of independent fragments, then that is how his mind will tend to operate, but if he can include verything coherently and harmoniously in an overall whole that is undivided, unbroken, and without a border (for every border is a division or break) then his mind will tend to move in a similar way, and from this will flow an orderly action within the whole.

Of course, as I have already indicated, our general world view is not the *only* factor that is important in this context. Attention must, indeed, be given to many other factors, such as emotions, physical activities, human relationships, social organizations, etc, but perhaps because we have at present no coherent world view, there is a widespread tendency to ignore the psychological and social importance of such questions almost altogether. My suggestion is that a proper world view, appropriate for its time, is generally one of the basic factors that is essential for harmony in the individual and in society as a whole.

In chapter 1 it is shown that science itself is demanding a new, non-fragmentary world view, in the sense that the present approach of analysis of the world into independently existent parts does not work very well in modern physics. It is shown that both in relativity theory and quantum theory, notions implying the undivided wholeness of the universe would provide a much more orderly way of considering the general nature of reality.

In chapter 2 we go into the role of language in bringing about fragmentation of thought. It is pointed out that the subject-verb-object structure of modern languages implies that all action arises in a separate subject, and acts either on a separate object, or else reflexively on itself. This pervasive structure leads in the whole of life to a function that divides the totality of existence into separate entities, which are considered to be essentially fixed and static in their nature. We then inquire whether it is possible to experiment with new language forms in which the basic role will be given to the verb rather than to the noun. Such forms will have as their content a series of actions that flow and merge into each other, without sharp separations or breaks. Thus, both in form and in content, the language will be in harmony with the unbroken flowing movement of existence as a whole.

What is proposed here is not a new language as such but, rather, a new *mode* of using the existing language – the *rheomode* (flowing mode). We develop such a mode as a form of experimentation with language, which is intended mainly to give insight into the fragmentary function of the common language rather than to provide a new way of speaking that can be used for practical communications.

In chapter 3 the same questions are considered within a different context. It begins with a discussion of how reality can be considered as in essence a set of forms in an underlying universal movement or process, and then asks how our knowledge can be considered in the same manner. Thus, the way could be opened for a world view in which consciousness and reality would not be fragmented from each other. This question is discussed at length and we arrive at the notion that our general world view is itself an overall movement of thought, which has to be viable in the sense that the totality of activities that flow out of it are generally in harmony, both in themselves and with regard to the whole of existence. Such harmony is seen to be possible only if the world view itself takes part in an unending process of development, evolution, and unfoldment, which fits as part of the universal process that is the ground of all existence.

The next three chapters are rather more technical and mathematical. However, large parts of them should be comprehensible to the non-technical reader, as the technical parts are not entirely necessary for comprehension, although they add significant content for those who can follow them.

Chapter 4 deals with hidden variables in the quantum theory.

The quantum theory is, at present, the most basic way available in physics for understanding the fundamental and universal laws relating to matter and its movement. As such, it must clearly be given serious consideration in any attempt to develop an overall world viewing.

The quantum theory, as it is now constituted, presents us with a very great challenge, if we are at all interested in such a venture, for in this theory there is no consistent notion at all of what the reality may be that underlies the universal constitution and structure of matter. Thus, if we try to use the prevailing world view based on the notion of particles, we discover that the 'particles' (such as electrons) can also manifest as waves, that they can move discontinuously, that there are no laws at all that apply in detail to the actual movements of individual particles and that only statistical predictions can be made about large aggregates of such particles. If on the other hand we apply the world view in which the universe is regarded as a continuous field, we find that this field must also be discontinuous, as well as particle-like, and that it is as undermined in its actual behaviour as is required in the particle view of relation as a whole.

It seems clear, then, that we are faced with deep and radical fragmentation, as well as thoroughgoing confusion, if we try to think of what could be the reality that is treated by our physical laws. At present physicists tend to avoid this issue by adopting the attitude that our overall views concerning the nature of reality are of little or no importance. All that counts in physical theory is supposed to be the development of mathematical equations that permit us to predict and control the behaviour of large statistical aggregates of particles. Such a goal is not regarded as merely for its pragmatic and technical utility: rather, it has become a presupposition of most work in modern physics that prediction and control of this kind is all that human knowledge is about.

This sort of presupposition is indeed in accord with the general spirit of our age, but it is my main proposal in this book that we cannot thus simply dispense with an overall world view. If we try to do so, we will find that we are left with whatever (generally inadequate) world views may happen to be at hand. Indeed, one finds that physicists are not actually able just to engage in calculations aimed at prediction and control: they do find it necessary to use images based on *some* kind of general notions concerning the nature of reality, such as 'the particles that are the building blocks of the universe'; but these images are now highly confused (e.g. these particles move discontinuously and are also waves). In short, we are here confronted with an example of how deep and strong is the need for *some* kind of notion of reality in our thinking, even if it be fragmentary and muddled.

My suggestion is that at each stage the proper order of operation of the mind requires an overall grasp of what is generally known, not only in formal, logical, mathematical terms, but also intuitively, in images, feelings, poetic usage of language, etc. (Perhaps we could say that this is what is involved in harmony between the 'left brain' and the 'right brain'.) This kind of overall way of thinking is not only a fertile source of new theoretical ideas: it is needed for the human mind to function in a generally harmonious way, which could in turn help to make possible an orderly and stable society. As indicated in the earlier chapters, however, this requires a continual flow and development of our general notions of reality.

Chapter 4 is then concerned with *making a beginning* in the process of developing a coherent view of what kind of reality might be the basis of the correct mathematical predictions achieved in the quantum theory. Such attempts have generally been received among the community of physicists in a somewhat confused way, for it is widely felt that if there is to be any general world view it should be taken as the 'received' and 'final' notion, concerning the nature of reality. But my attitude has, from the beginning, been that our notions concerning cosmology and the general nature of reality are in a continuous process of development, and that one may have to start with ideas that are merely some sort of improvement over what has thus far been available, and to go on from there to ideas that are better. Chapter 4 presents the real and severe problems that confront any attempt to provide a consistent notion of 'quantum-mechanical reality', and indicates a certain preliminary approach to a solution of these problems in terms of hidden variables.

In chapter 5 a different approach to the same problems is explored. This is an inquiry into our basic notions of order. Order in its totality is evidently ultimately undefinable, in the sense that it pervades everything that we are and do (language, thought, feeling, sensation, physical action, the arts, practical activity, etc.). However, in physics the basic order has for centuries been that of the Cartesian rectilinear grid (extended slightly in the theory of relativity to the curvilinear grid). Physics has had an enormous development during this time, with the appearance of many radically new features, but the basic order has remained essentially unchanged.

The Cartesian order is suitable for analysis of the world into separately existent parts (e.g. particles or field elements). In this chapter, however, we look into the nature of order with greater generality and depth, and discover that both in relativity and in quantum theory the Cartesian order is leading to serious contradictions and confusion. This is because both theories imply that the actual state of affairs is unbroken wholeness of the universe, rather than analysis into independent parts. Nevertheless, the two theories differ radically in their detailed notions of order. Thus, in relativity, movement is continuous, causally determinate and well defined, while in quantum mechanics it is discontinuous, not causally determinate and not well defined. Each theory is committed to its own notions of essentially static and fragmentary modes of existence (relativity to that of separate events, connectable by signals, and quantum mechanics to a well-defined quantum state). One thus sees that a new kind of theory is needed, which drops these basic commitments and at most recovers some essential features of the older theories as abstract forms, derived from a deeper reality in which what prevails is unbroken wholeness.

In chapter 6 we go further to begin a more concrete development of a new notion of order, that may be appropriate to a universe of unbroken wholeness. This is the *implicate* or *enfolded* order. In the enfolded order, space and time are no longer the dominant factors determining the relationships of dependence or independence of different elements. Rather, an entirely different sort of basic connection of elements is possible, from which our ordinary notions of space and time, along with those of separately existent material particles, are abstracted as forms derived from the deeper order. These ordinary notions in fact appear in what is called the *explicate* or *unfolded* order, which is a special and distinguished form, contained within the general totality of all the implicate orders.

In chapter 6 the implicate order is introduced in a general way, and discussed mathematically in an appendix. The seventh and last chapter, however, is a more developed (though non-technical) presentation of the implicate order, along with its relationship to consciousness. This leads to an indication of some lines along which it may be possible to meet the urgent challenge to develop a cosmology and set of general notions concerning the nature of reality that are proper to our time.

Finally, it is hoped that the presentation of the material in these essays may help to convey to the reader how the subject itself has actually unfolded, so that the form of the book is, as it were, an example of what may be meant by the content.

1

Fragmentation and wholeness

The title of this chapter is 'Fragmentation and wholeness'. It is especially important to consider this question today, for fragmentation is now very widespread, not only throughout society, but also in each individual; and this is leading to a kind of general confusion of the mind, which creates an endless series of problems and interferes with our clarity of perception so seriously as to prevent us from being able to solve most of them.

Thus art, science, technology, and human work in general, are divided up into specialities, each considered to be separate in essence from the others. Becoming dissatisfied with this state of affairs, men have set up further interdisciplinary subjects, which were intended to unite these specialities, but these new subjects have ultimately served mainly to add further separate fragments. Then, society as a whole has developed in such a way that it is broken up into separate nations and different religious, political, economic, racial groups, etc. Man's natural environment has correspondingly been seen as an aggregate of separately existent parts, to be exploited by different groups of people. Similarly, each individual human being has been fragmented into a large number of separate and conflicting compartments, according to his different desires, aims, ambitions, loyalties, psychological characteristics, etc., to such an extent that it is generally accepted that some degree of neurosis is inevitable, while many individuals going beyond the 'normal' limits of fragmentation are classified as paranoid, schizoid, psychotic, etc.

The notion that all these fragments are separately existent is evidently an illusion, and this illusion cannot do other than lead to endless conflict and confusion. Indeed, the attempt to live according to the notion that the fragments are really separate is, in essence, what has led to the growing series of extremely urgent crises that is confronting us today. Thus, as is now well known, this way of life has brought about pollution, destruction of the balance of nature, over-population, world-wide economic and political disorder, and the creation of an overall environment that is neither physically nor mentally healthy for most of the people who have to live in it. Individually there has developed a widespread feeling of helplessness and despair, in the face of what seems to be an overwhelming mass of disparate social forces, going beyond the control and even the comprehension of the human beings who are caught up in it.

Indeed, to some extent, it has always been both necessary and proper for man, in his thinking, to divide things up, and to separate them, so as to reduce his problems to manageable proportions; for evidently, if in our practical technical work we tried to deal with the whole of reality all at once, we would be swamped. So, in certain ways, the creation of special subjects of study and the division of labour was an important step forward. Even earlier, man's first realization that he was not identical with nature was also a crucial step, because it made possible a kind of autonomy in his thinking, which allowed him to go beyond the immediately given limits of nature, first in his imagination and ultimately in his practical work.

Nevertheless, this sort of ability of man to separate himself from his environment and to divide and apportion things, ultimately led to a wide range of negative and destructive results, because man lost awareness of what he was doing and thus extended the process of division beyond the limits within which it works properly. In essence, the process of division is a way of *thinking about things* that is convenient and useful mainly in the domain of practical, technical and functional activities (e.g., to divide up an area of land into different fields where various crops are to be grown). However, when this mode of thought is applied more broadly to man's notion of himself and the whole world in which he lives (i.e. to his self-world view), then man ceases to regard the resulting divisions as merely useful or convenient and begins to see and experience himself and his world as actually constituted of separately existent fragments. Being guided by a fragmentary self-world view, man then acts in such a way as to try to break himself and the world up, so that all seems to correspond to his way of thinking. Man thus obtains an apparent proof of the correctness of his fragmentary self-world view though, of course, he overlooks the fact that it is he himself, acting according to his mode of thought, who has brought about the fragmentation that now seems to have an autonomous existence, independent of his will and of his desire.

Men have been aware from time immemorial of this state of apparently autonomously existent fragmentation and have often projected myths of a yet earlier 'golden age', before the split between man and nature and between man and man had yet taken place. Indeed, man has always been seeking wholeness – mental, physical, social, individual.

It is instructive to consider that the word 'health' in English is based on an Anglo-Saxon word 'hale' meaning 'whole': that is, to be healthy is to be whole, which is, I think, roughly the equivalent of the Hebrew 'shalem'. Likewise, the English 'holy' is based on the same root as 'whole'. All of this indicates that man has sensed always that wholeness or integrity is an absolute necessity to make life worth living. Yet, over the ages, he has generally lived in fragmentation.

Surely, the question of why all this has come about requires careful attention and serious consideration.

In this chapter, attention will be focused on the subtle but crucial role of our general forms of thinking in sustaining fragmentation and in defeating our deepest urges toward wholeness or integrity. In order to give the discussion a concrete content we shall to some extent talk in terms of current scientific research, which is a field that is relatively familiar to me (though, of course, the overall significance of the questions under discussion will also be kept in mind).

What will be emphasized, first of all in scientific research and later in a more general context, is that fragmentation is continually being brought about by the almost universal habit of taking the content of our thought for 'a description of the world as it is'. Or we could say that, in this habit, our thought is regarded as in direct correspondence with objective reality. Since our thought is pervaded with differences and distinctions, it follows that such a habit leads us to look on these as real divisions, so that the world is then seen and experienced as actually broken up into fragments.

The relationship between thought and reality that this thought is about is in fact far more complex than that of a mere correspondence. Thus, in scientific research, a great deal of our thinking is in terms of *theories*. The word 'theory' derives from the Greek 'theoria', which has the same root as 'theatre', in a word meaning 'to view' or 'to make a spectacle'. Thus, it might be said that a theory is primarily a form of *insight*, i.e. a way of looking at the world, and not a form of *knowledge* of how the world is.

In ancient times, for example, men had the theory that celestial matter was fundamentally different from earthly matter and that it was natural for earthly objects to fall while it was natural for celestial objects, such as the moon, to remain up in the sky. With the coming of the modern era, however, scientists began to develop the viewpoint that there was no essential difference between earthly matter and celestial matter. This implied, of course, that heavenly objects, such as the moon, ought to fall, but for a long time men did not notice this implication. In a sudden flash of insight Newton then *saw* that as the apple falls so does the moon, and so indeed do all objects. Thus, he was led to the theory of universal gravitation, in which all objects were seen as falling toward various centres (e.g. the earth, the sun, the planets, etc.). This constituted a new way of *looking* at the heavens, in which the movements of the planets were no longer seen through the ancient notion of an essential difference between heavenly and earthly matter. Rather, one considered these movements in terms of rates of fall of all matter, heavenly and earthly, toward various centres, and when something was seen not to be accounted for in this way, one looked for and often discovered new and as yet unseen planets toward which celestial objects were falling (thus demonstrating the relevance of this way of looking).

The Newtonian form of insight worked very well for several centuries but ultimately (like the ancient Greek insights that came before) it led to unclear results when extended into new domains. In these new domains, new forms of insight were developed (the theory of relativity and the quantum theory). These gave a radically different picture of the world from that of Newton (though the latter was, of course, found to be still valid in a limited domain). If we supposed that theories gave true knowledge, corresponding to 'reality as it is', then we would have have to conclude that Newtonian theory was true until around 1900, after which it suddenly became false, while relativity and quantum theory suddenly became the truth. Such an absurd conclusion does not arise, however, if we say that all theories are insights, which are neither true nor false but, rather, clear in certain domains, and unclear when extended beyond these domains. This means, however, that we do not equate theories with hypotheses. As the Greek root of the word indicates, a hypothesis is a supposition, that is, an idea that is 'put under' our reasoning, as a provisional base, which is to be tested experimentally for its truth or falsity. As is now well known, however, there can be no *conclusive* experimental proof of the truth or falsity of a *general* hypothesis which aims to cover the whole of reality. Rather, one finds (e.g., as in the case of the Ptolemaic epicycles or of the failure of Newtonian concepts just before the advent of relativity and quantum theory) that older theories become more and more unclear when one tries to use them to obtain insight into new domains. Careful attention to how this happens is then generally the main clue toward new theories, that constitute further new forms of insight.

So, instead of supposing that older theories are falsified at a certain point in time, we merely say that man is continually developing new forms of insight, which are clear up to a point and then tend to become unclear. In this activity, there is evidently no reason to suppose that there is or will be a final form of insight (corresponding to absolute truth) or even a steady series of approximations to this. Rather, in the nature of the case, one may expect the unending development of new forms of insight (which will, however, assimilate certain key features of the older forms as simplifications, in the way that relativity theory does with Newtonian theory). As pointed out earlier, however, this means that our theories are to be regarded primarily as ways of looking at the world as a whole (i.e. world views) rather than as 'absolutely true knowledge of how things are' (or as a steady approach toward the latter).

When we look at the world through our theoretical insights, the factual knowledge that we obtain will evidently be shaped and formed by our theories. For example, in ancient times the fact about the motions of the planets was described in terms of the Ptolemaic idea of epicycles (circles superimposed on circles). In Newton's time, this fact was described in terms of precisely determined planetary orbits, analysed through rates of fall toward various centres. Later came the fact as seen relativistically according to Einstein's concepts of space and time. Still later, a very different sort of fact was specified in terms of the quantum theory (which gives in general only a statistical fact). In biology, the fact is now described in terms of the theory of evolution, but in earlier times it was expressed in terms of fixed species of living beings.

More generally, then, given perception and action, our theoretical insights provide the main source of organization of our factual knowledge. Indeed, our overall experience is shaped in this way. As seems to have been first pointed out by Kant, all experience is organized according to the categories of our thought, i.e., on our ways of thinking about space, time, matter, substance, causality, contingency, necessity, universality, particularity, etc. It can be said that these categories are general forms of insight or ways of looking at everything, so that in a certain sense, they are a kind of theory (but, of course, this level of theory must have developed very early in man's evolution).

Clarity of perception and thought evidently requires that we be generally aware of how our experience is shaped by the insight (clear or confused) provided by the theories that are implicit or explicit in our general ways of thinking. To this end, it is useful to emphasize that experience and knowledge are one process, rather than to think that our knowledge is *about* some sort of separate experience. We can refer to this one process as experience-knowledge (the hyphen indicating that these are two inseparable aspects of one whole movement.)

Now, if we are not aware that our theories are ever-changing forms of insight, giving shape and form to experience in general, our vision will be limited. One could put it like this: experience with nature is very much like experience with human beings. If one approaches another man with a fixed 'theory' about him as an 'enemy' against whom one must defend oneself, he will respond similarly, and thus one's 'theory' will apparently be confirmed by experience. Similarly, nature will respond in accordance with the theory with which it is approached. Thus, in ancient times, men thought plagues were inevitable, and this thought helped make them behave in such a way as to propagate the conditions responsible for their spread. With modern scientific forms of insights man's behaviour is such that he ceases the insanitary modes of life responsible for spreading plagues and thus they are no longer inevitable.

What prevents theoretical insights from going beyond existing limitations and changing to meet new facts is just the belief that theories give true knowledge of reality (which implies, of course, that they need never change). Although our modern way of thinking has, of course, changed a great deal relative to the ancient one, the two have had one key feature in common: i.e. they are both generally 'blinkered' by the notion that theories give true knowledge about 'reality as it is'. Thus, both are led to confuse the forms and shapes induced in our perceptions by theoretical insight with a reality independent of our thought and our way of looking. This confusion is of crucial significance, since it leads us to approach nature, society, and the individual in terms of more or less fixed and limited forms of thought, and thus, apparently, to keep on confirming the limitations of these forms of thought in experience.

This sort of unending confirmation of limitations in our modes of thinking is particularly significant with regard to fragmentation, for as pointed out earlier, every form of theoretical insight introduces its own essential differences and distinction (e.g., in ancient times an essential distinction was between heavenly and earthly matter, while in Newtonian theory it was essential to distinguish the centres toward which all matter was falling). If we regard these differences and distinctions as ways of looking, as guides to perception, this does not imply that they denote separately existent substances or entities.

On the other hand, if we regard our theories as 'direct descriptions of reality as it is', then we will inevitably treat these differences and distinction as divisions, implying separate existence of the various elementary terms appearing in the theory. We will thus be led to the illusion that the world is actually constituted of separate fragments and, as has already been indicated, this will cause us to act in such a way that we do in fact produce the very fragmentation implied in our attitude to the theory.

It is important to give some emphasis to this point. For example, some might say: 'Fragmentation of cities, religions, political systems, conflict in the form of wars, general violence, fratricide, etc., are the reality. Wholeness is only an ideal, toward which we should perhaps strive.' But this is not what is being said here. Rather, what should be said is that wholeness is what is real, and that fragmentation is the response of this whole to man's action, guided by illusory perception, which is shaped by fragmentary thought. In other words, it is just because reality is whole that man, with his fragmentary approach, will inevitably be answered with a correspondingly fragmentary response. So what is needed is for man to give attention to his habit of fragmentary thought, to be aware of it, and thus bring it to an end. Man's approach to reality may then be whole, and so the response will be whole.

For this to happen, however, it is crucial that man be aware of the activity of his thought *as such*; i.e. as a form of insight, a way of looking, rather than as a 'true copy of reality as it is'.

It is clear that we may have any number of different kinds of insights. What is called for is not an *integration* of thought, or a kind of imposed unity, for any such imposed point of view would itself be merely another fragment. Rather, all our different ways of thinking are to be considered as different ways of looking at the one reality, each with some domain in which it is clear and adequate. One may indeed compare a theory to a particular view of some object. Each view gives only an appearance of the object in some aspect. The whole object is not perceived in any one view but, rather, it is grasped only *implicitly* as that single reality which is shown in all these views. When we deeply understand that our theories also work in this way, then we will not fall into the habit of seeing reality and acting toward it as if it were constituted of separately existent fragments, corresponding to how it appears in our thought and in our imagination, when we take our theories to be 'direct descriptions of reality as it is'.

Beyond a general awareness of the role of theories as indicated above, what is needed is to give special attention to those theories that contribute to the expression of our overall self-world views. For, to a considerable extent, it is in these world views that our general notions of the nature of reality and of the relationship between our thought and reality are implicity or explicitly formed. In this respect, the general theories of physics play an important part, because they are regarded as dealing with the universal nature of the matter out of which all is constituted, and the space and time in terms of which all material movement is described.

Consider, for example, the atomic theory, which was first proposed by Democritus more than 2,000 years ago. In essence, this theory leads us to look at the world as constituted of atoms, moving in the void. The ever-changing forms and characteristics of large-scale objects are now seen as the results of changing arrangements of the moving atoms. Evidently, this view was, in certain ways, an important mode of realization of wholeness, for it enabled men to understand the enormous variety of the whole world in terms of the movements of one single set of basic constituents, through a single void that permeates the whole of existence. Nevertheless, as the atomic theory developed, it ultimately became a major support for a fragmentary approach to reality. For it ceased to be regarded as an insight, a way of looking, and men regarded instead as an absolute truth the notion that the whole of reality is actually constituted of nothing but 'atomic building blocks', all working together more or less mechanically.

Of course, to take any physical theory as an absolute truth must tend to fix the general forms of thought in physics and thus to contribute to fragmentation. Beyond this, however, the particular content of the atomic theory was such as to be especially conducive to fragmentation, for it was implicit in this content that the entire world of nature, along with the human being, including his brain, his nervous system, his mind, etc., could in principle be understood completely in terms of structures and functions of aggregates of separately existent atoms. The fact that in man's experiments and general experience this atomic view was confirmed was, of course, then taken as proof of the correctness and indeed the universal truth of this notion. Thus almost the whole weight of science was put behind the fragmentary approach to reality.

It is important to point out, however, that (as usually happens in such cases) the experimental confirmation of the atomic point of view is limited. Indeed, in the domains covered by quantum theory and relativity, the notion of atomism leads to confused questions, which indicate the need for new forms of insight, as different from atomism as the latter is from theories that came before it.

Thus, the quantum theory shows that the attempt to describe and follow an atomic particle in precise detail has little meaning. (Further detail on this point is given in chapter 5). The notion of an atomic path has only a limited domain of applicability. In a more detailed description the atom is, in many ways, seen to behave as much like a wave as a particle. It can perhaps best be regarded as a poorly defined cloud, dependent for its particular form on the whole environment, including the observing instrument. Thus, one can no longer maintain the division between the observer and observed (which is implicit in the atomistic view that regards each of these as separate aggregates of atoms). Rather, both observer and observed are merging and interpenetrating aspects of one whole reality, which is indivisible and unanalysable.

Relativity leads us to a way of looking at the world that is similar to the above in certain key respects (See chapter 5 for more detail on this point). From the fact that in Einstein's point of view no signal faster than light is possible, it follows that the concept of a rigid body breaks down. But this concept is crucial in the classical atomic theory, for in this theory the ultimate constituents of the universe have to be small indivisible objects, and this is possible only if each part of such an object is bound rigidly to all other parts. What is needed in a relativistic theory is to give up altogether the notion that the world is constituted of basic objects or 'building blocks'. Rather, one has to view the world in terms of universal flux of events and processes. Thus, as indicated by A and B in figure 1.1, instead of thinking of a particle, one is to think of a 'world tube'.

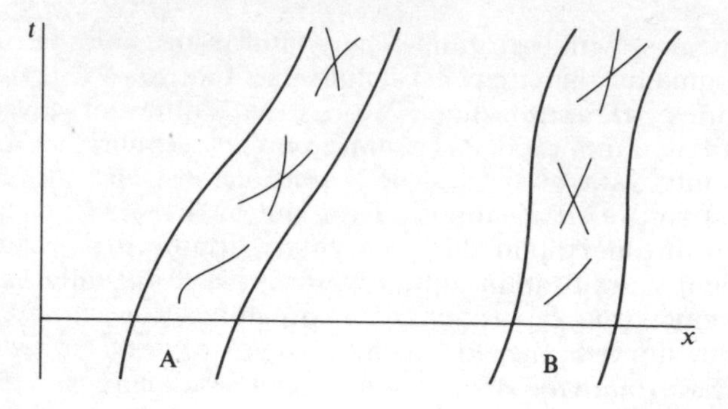

Figure 1.1

This world tube represents an infinitely complex process of a structure in movement and development which is centred in a region indicated by the boundaries of the tube. However, even outside the tube, each 'particle' has a field that extends through space and merges with the fields of other particles.

A more vivid image of the sort of thing that is meant is afforded by considering wave forms as vortex structures in a flowing stream. As shown in figure 1.2, two vortices correspond to stable patterns of flow of the fluid, centred more or less at A and B. Evidently, the two vortices are to be considered as abstractions, made to stand out in our perception by our way of thinking. Actually, of course, the two abstracted flow patterns merge and unite, in one whole movement of the flowing stream. There is no sharp division between them, nor are they to be regarded as separately or independently existent entities.

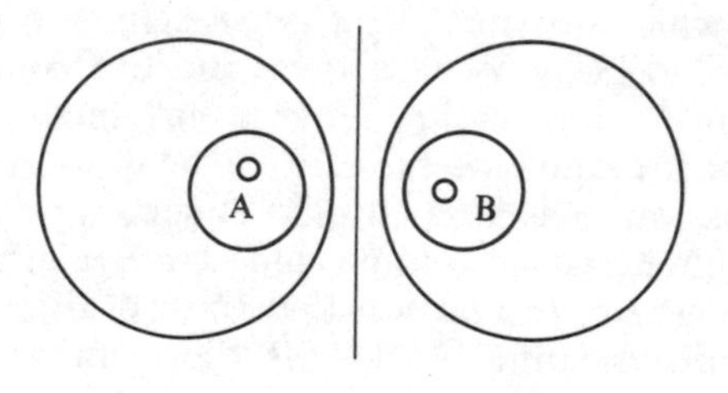

Figure 1.2

Relativity theory calls for this sort of way of looking at the atomic particles, which constitute all matter, including of course human beings, with their brains, nervous systems, and the observing instruments that they have built and that they use in their laboratories. So, approaching the question in different ways, relativity and quantum theory agree, in that they both imply the need to look on the world as an *undivided whole*, in which all parts of the universe, including the observer and his instruments, merge and unite in one totality. In this totality, the atomistic form of insight is a simplification and an abstraction, valid only in some limited context.

The new form of insight can perhaps best be called *Undivided Wholeness in Flowing Movement*. This view implies that flow is, in some sense, prior to that of the 'things' that can be seen to form and dissolve in this flow. One can perhaps illustrate what is meant here by considering the 'stream of consciousness'. This flux of awareness is not precisely definable, and yet it is evidently prior to the definable forms of thoughts and ideas which can be seen to form and dissolve in the flux, like ripples, waves and vortices in a flowing stream. As happens with such patterns of movement in a stream some thoughts recur and persist in a more or less stable way, while others are evanescent.

The proposal for a new general form of insight is that all matter is of this nature: That is, there is a universal flux that cannot be defined explicitly but which can be known only implicitly, as indicated by the explicitly definable forms and shapes, some stable and some unstable, that can be abstracted from the universal flux. In this flow, mind and matter are not separate substances. Rather, they are different aspects of one whole and unbroken movement. In this way, we are able to look on all aspects of existence as not divided from each other, and thus we can bring to an end the fragmentation implicit in the current attitude toward the atomic point of view, which leads us to divide everything from everything in a thoroughgoing way. Nevertheless, we can comprehend that aspect of atomism which still provides a correct and valid form of insight; i.e. that in spite of the undivided wholeness in flowing movement, the various patterns that can be abstracted from it have a certain relative autonomy and stability, which is indeed provided for by the universal law of the flowing movement. Now, however, we have the limits of this autonomy and stability sharply in mind.

Thus we can, in specified contexts, adopt other various forms of insight that enable us to simplify certain things and to treat them momentarily and for certain limited purposes as if they were autonomous and stable, as well as perhaps separately existent.

Yet we do not have to fall into the trap of looking at ourselves and at the whole world in this way. Thus our thought need no longer lead to the illusion that reality actually is of fragmentary nature, and to the corresponding fragmentary actions that arise out of perception clouded by such illusion.

The point of view discussed above is similar, in certain key ways, to that held by some of the Ancient Greeks. This similarity can be brought out by considering Aristotle's notion of causality. Aristotle distinguished four kinds of causes:

Material
Efficient
Formal
Final

A good example in terms of which this distinction can be understood is obtained by considering something living, such as a tree or an animal. The material cause is then just the matter in which all the other causes operate and out of which the thing is constituted. Thus, in the case of a plant, the material cause is the soil, air, water and sunlight, constituting the substance of the plant. The efficient cause is some action, external to the thing under discussion, which allows the whole process to get under way. In the case of a tree, for example, the planting of the seed could be taken as the efficient cause.

It is of crucial significance in this context to understand what was meant by formal cause. Unfortunately, in its modern connotation, the word 'formal' tends to refer to an outward form that is not very significant (e.g. as in 'formal dress' or 'a mere formality'). However, in the Ancient Greek philosophy, the word *form* meant, in the first instance, an inner *forming activity* which is the cause of the growth of things, and of the development and differentiation of their various essential forms. For example, in the case of an oak tree, what is indicated by the term 'formal cause' is the whole inner movement of sap, cell growth, articulation of branches, leaves, etc., which is characteristic of that kind of tree and different from that taking place in other kinds of trees. In more modern language, it would be better to describe this as *formative cause*, to emphasize that what is involved is not a mere form imposed from without, but rather *an ordered and structured inner movement that is essential to what things are.*

Any such formative cause must evidently have an end or product, which is at least implicit. Thus, it is not possible to refer to the inner movement from the acorn giving rise to an oak tree, without simultaneously referring to the oak tree that is going to result from this movement. So formative cause always implies final cause.

Of course, we also know final cause as *design*, consciously held in mind through thought (this notion being extended to God, who was regarded as having created the universe according to some grand design). Design is, however, only a special case of final cause. For example, men often aim toward certain ends in their thoughts but what actually emerges from their actions is generally something different from what was in their design, something that was, however, *implicit* in what they were doing, though not consciously perceived by those who took part.

In the ancient view, the notion of formative cause was considered to be of essentially the same nature for the mind as it was for life and for the cosmos as a whole. Indeed, Aristotle considered the universe as a single organism in which each part grows and develops in its relationship to the whole and in which it has its proper place and function. With regard to the mind, we can understand this sort of notion in more modern terms by turning our attention to the flowing movement of awareness. As indicated earlier, one can, in the first instance, discern various thought patterns in this flow. These follow on each other relatively mechanically, through association determined by habit and conditioning. Evidently, such associative changes are external to the inner structure of the thoughts in question, so that these changes act like a series of efficient causes. However, to see the *reason* for something is not a mechanical activity of this nature: Rather, one is aware of each aspect as assimilated within a single whole, all of whose parts are inwardly related (as are, for example, the organs of the body). Here, one has to emphasize that the act of reason is essentially a kind of perception through the mind, similar in certain ways to artistic perception, and not merely the associative repetition of reasons that are already known. Thus, one may be puzzled by a wide range of factors, things that do not fit together, until suddenly there is a flash of understanding, and therefore one sees how all these factors are related as aspects of one totality (e.g. consider Newton's insight into universal gravitation). Such acts of perception cannot properly be given a detailed analysis or description. Rather, they are to be considered as aspects of the *forming* activity of the mind. A particular structure of concepts is then the *product* of this activity, and these products are what are linked by the series of efficient causes that operate in ordinary associative thinking – and as pointed out earlier, in this view, one regards the forming activity as primary in nature as it is in the mind, so that the product forms in nature are also what are linked by efficient causes.

Evidently, the notion of formative cause is relevant to the view of undivided wholeness in flowing movement, which has been seen to be implied in modern developments in physics, notably relativity theory and quantum theory. Thus, as has been pointed out, each relatively autonomous and stable structure (e.g., an atomic particle) is to be understood not as something independently and permanently existent but rather as a product that has been formed in the whole flowing movement and that will ultimately dissolve back into this movement. How it forms and maintains itself, then, depends on its place and function in the whole. So, we see that certain developments in modern physics imply a sort of insight into nature that is in respect to the notions of formative and final cause, essentially similar to ways of looking that were common in earlier times.

Nevertheless, in most of the work that is being done in physics today the notions of formative and final cause are not regarded as having primary significance. Rather, law is still generally conceived as a self-determined system of efficient causes, operating in an ultimate set of material constituents of the universe (e.g. elementary particles subject to forces of interaction between them). These constituents are not regarded as formed in an overall process, and thus they are not considered to be anything like organs adapted to their place and function in the whole (i.e. to the ends which they would serve in this whole). Rather, they tend to be conceived as separately existent mechanical elements of a fixed nature.

The prevailing trend in modern physics is thus much against any sort of view giving primacy to formative activity in undivided wholeness of flowing movement. Indeed, those aspects of relativity theory and quantum theory which do suggest the need for such a view tend to be de-emphasized and in fact hardly noticed by most physicists, because they are regarded largely as features of the mathematical calculus and not as indications of the real nature of things. When it comes to the informal language and mode of thought in physics, which infuses the imagination and provokes the sense of what is real and substantial, most physicists still speak and think, with an utter conviction of truth, in terms of the traditional atomistic notion that the universe is constituted of elementary particles which are 'basic building blocks' out of which everything is made. In other sciences, such as biology, the strength of this conviction is even greater, because among workers in these fields there is little awareness of the revolutionary character of development in modern physics. For example, modern molecular biologists generally believe that the whole of life and mind can ultimately be understood in more or less mechanical terms, through some kind of extension of the work that has been done on the structure and function of DNA molecules. A similar trend has already begun to dominate in psychology. Thus we arrive at the very odd result that in the study of life and mind, which are just the fields in which formative cause acting in undivided and unbroken flowing movement is most evident to experience and observation, there is now the strongest belief in the fragmentary atomistic approach to reality.

Of course, the prevailing tendency in science to think and perceive in terms of a fragmentary self-world view is part of a larger movement that has been developing over the ages and that pervades almost the whole of our society today: but, in turn, such a way of thinking and looking in scientific research tends very strongly to re-enforce the general fragmentary approach because it gives men a picture of the whole world as constituted of nothing but an aggregate of separately existent 'atomic building blocks', and provides experimental evidence from which is drawn the conclusion that this view is necessary and inevitable. In this way, people are led to feel that fragmentation is nothing but an expression of 'the way everything really is' and that anything else is impossible. So there is very little disposition to look for evidence to the contrary. Indeed, as has already been pointed out, even when such evidence does arise, as in modern physics, the general tendency is to minimize its significance or even to ignore it altogether. One might in fact go so far as to say that in the present state of society, and in the present general mode of teaching science, which is a manifestation of this state of society, a kind of prejudice in favour of a fragmentary self-world view is fostered and transmitted (to some extent explicitly and consciously but mainly in an implicit and unconscious manner).

As has been indicated, however, men who are guided by such a fragmentary self-world view cannot, in the long run, do other than to try in their actions to break themselves and the world into pieces, corresponding to their general mode of thinking. Since, in the first instance, fragmentation is an attempt to extend the analysis of the world into separate parts beyond the domain in which to do this is appropriate, it is in effect an attempt to divide what is really indivisible. In the next step such an attempt will lead us also to try to unite what is not really unitable. This can be seen especially clearly in terms of groupings of people in society (political, economic, religious, etc.). The very act of forming such a group tends to create a sense of division and separation of the members from the rest of the world but, because the members are really connected with the whole, this cannot work. Each member has in fact a somewhat different connection, and sooner or later this shows itself as a difference between him and other members of the group. Whenever men divide themselves from the whole of society and attempt to unite by identification within a group, it is clear that the group must eventually develop internal strife, which leads to a breakdown of its unity. Likewise when men try to separate some aspect of nature in their practical, technical work, a similar state of contradiction and disunity will develop. The same sort of thing will happen to the individual when he tries to separate himself from society. True unity in the individual and between man and nature, as well as between man and man, can arise only in a form of action that does not attempt to fragment the whole of reality.

Our fragmentary way of thinking, looking, and acting, evidently has implications in every aspect of human life. That is to say, by a rather interesting sort of irony, fragmentation seems to be the one thing in our way of life which is universal, which works through the whole without boundary or limit. This comes about because the roots of fragmentation are very deep and pervasive. As pointed out, we try to divide what is one and indivisible, and this implies that in the next step we will try to identify what is different.

So fragmentation is in essence a confusion around the question of difference and sameness (or one-ness), but the clear perception of these categories is necessary in every phase of life. *To be confused about what is different and what is not, is to be confused about everything.* Thus, it is not an accident that our fragmentary form of thought is leading to such a widespread range of crises, social, political, economic, ecological, psychological, etc., in the individual and in society as a whole. Such a mode of thought implies unending development of chaotic and meaningless conflict, in which the energies of all tend to be lost by movements that are antagonistic or else at cross-purposes.

Evidently, it is important and indeed extremely urgent to clear up this deep and pervasive kind of confusion that penetrates the whole of our lives. What is the use of attempts at social, political, economic or other action if the mind is caught up in a confused movement in which it is generally differentiating what is not different and identifying what is not identical? Such action will be at best ineffective and at worst really destructive.

Nor will it be useful to try to impose some fixed kind of integrating or unifying 'holistic' principle on our self-world view, for, as indicated earlier, any form of fixed self-world view implies that we are no longer treating our theories as insights or ways of looking but, rather, as 'absolutely true knowledge of things as they really are'. So, whether we like it or not, the distinctions that are inevitably present in every theory, even an 'holistic' one, will be falsely treated as divisions, implying separate existence of the terms that are distinguished (so that, correspondingly, what is not distinguished in this way will be falsely treated as absolutely identical).

We have thus to be alert to give careful attention and serious consideration to the fact that our theories are not 'descriptions of reality as it is' but, rather, ever-changing forms of insight, which can point to or indicate a reality that is implicit and not describable or specifiable in its totality. This need for being thus watchful holds even for what is being said here in this chapter, in the sense that this is not to be regarded as 'absolutely true knowledge of the nature of fragmentations and wholeness'. Rather, it too is *a theory* that gives insight into this question. It is up to the reader to see for himself whether the insight is clear or unclear and what are the limits of its validity.

What, then, can be done to end the prevailing state of fragmentation? At first sight this may seem to be a reasonable question but a closer examination leads one to ask whether it is in fact a reasonable question, for one can see that this question has presuppositions that are not clear.

Generally speaking, if one asks how one can solve some technical problem, for example, it is presupposed that while we begin not knowing the answer, our minds are nevertheless clear enough to discover an answer, or at least to recognize someone else's discovery of an answer. But if our whole way of thinking is penetrated by fragmentation, this implies that we are not capable of this, for fragmentary perception is in essence a largely unconscious habit of confusion around the question of what is different and what is not. So, in the very act in which we try to discover what to do about fragmentation, we will go on with this habit and thus we will tend to introduce yet further forms of fragmentation.

This does not necessarily mean, of course, that there is no way out at all, but it does mean that we have to give pause so that we do not go with our habitual fragmentary ways of thinking as we seek solutions that are ready to hand. The question of fragmentation and wholeness is a subtle and difficult one, more subtle and difficult than those which lead to fundamentally new discoveries in science. To ask how to end fragmentation and to expect an answer in a few minutes makes even less sense than to ask how to develop a theory as new as Einstein's was when he was working on it, and to expect to be told what to do in terms of some programme, expressed in terms of formulae or recipes.

One of the most difficult and subtle points about this question is just to clarify what is to be meant by the relationship between the content of thought and the process of thinking which produces this content. A major source of fragmentation is indeed the generally accepted presupposition that the process of thought is sufficiently separate from and independent of its content, to allow us generally to carry out clear, orderly, rational thinking, which can properly judge this content as correct or incorrect, rational or irrational, fragmentary or whole, etc. Actually, as has been seen, the fragmentation involved in a self-world view is not only in the content of thought, but in the general activity of the person who is 'doing the thinking', and thus, it is as much in the process of thinking as it is in the content. Indeed, content and process are not two separately existent things, but, rather, they are two aspects of views of one whole movement. Thus fragmentary content and fragmentary process have to come to an end *together*.

What we have to deal with here is a one-ness of the thinking process and its content, similar in key ways to the one-ness of observer and observed; that has been discussed in connection with relativity theory and quantum theory. Questions of this nature cannot be met properly while we are caught up, consciously or unconsciously, in a mode of thought which attempts to analyse itself in terms of a presumed separation between the process of thinking and the content of thought that is its product. By accepting such a presumption we are led, in the next step, to seek some fantasy of action through efficient causes, that would end the fragmentation in the content while leaving the fragmentation in the actual process of thinking untouched. What is needed, however, is somehow to grasp the overall *formative cause* of fragmentation, in which content and actual process are seen together, in their wholeness.

One might here consider the image of a turbulent mass of vortices in a stream. The structure and distribution of vortices, which constitute a sort of content of the description of the movement, are not separate from the formative activity of the flowing stream, which creates, maintains, and ultimately dissolves the totality of vortex structures. So to try to eliminate the vortices without changing the formative activity of the stream would evidently be absurd. Once our perception is guided by the proper insight into the significance of the whole movement, we will evidently not be disposed to try such a futile approach. Rather, we will look at the whole situation, and be attentive and alert to learn about it, and thus to discover what is really an appropriate sort of action relevant to this whole, for bringing the turbulent structure of vortices to an end. Similarly, when we really grasp the truth of the one-ness of the thinking process that we are actually carrying out, and the content of thought that is the product of this process, then such insight will enable us to observe, to look, to learn about the whole movement of thought and thus to discover an action relevant to this whole, that will end the 'turbulence' of movement which is the essence of fragmentation in every phase of life.

Of course, such learning and discovery will require a great deal of careful attention and hard work. We are ready to give such attention and work in a wide range of fields, scientific, economic, social, political, etc. As yet, however, little or none of this has gone into the creation of insight into the process of thought, on the clarity of which the value of all else depends. What is primarily needed is a growing realization of the extremely great danger of going on with a fragmentary process of thought. Such a realization would give the inquiry into how thought actually operates that sense of urgency and energy, required to meet the true magnitude of the difficulties with which fragmentation is now confronting us.

APPENDIX: RÉSUMÉ OF DISCUSSION ON WESTERN AND EASTERN FORMS OF INSIGHT INTO WHOLENESS

In the very early phases of the development of civilization, man's views were essentially of wholeness rather than of fragmentation. In the East (especially in India) such views still survive, in the sense that philosophy and religion emphasize wholeness and imply the futility of analysis of the world into parts. Why, then, do we not drop our fragmentary Western approach and adopt these Eastern notions which include not only a self-world view that denies division and fragmentation but also techniques of meditation that lead the whole process of mental operation non-verbally to the sort of quiet state of orderly and smooth flow needed to end fragmentation both in the actual process of thought and in its content?

To answer such a question, it is useful to begin by going into the difference between Western and Eastern notions of measure. Now, in the West the notion of measure has, from very early times, played a key role in determining the general self-world view and the way of life implicit in such a view. Thus among the Ancient Greeks, from whom we derive a large part of our fundamental notions (by way of the Romans), to keep everything in its right measure was regarded as one of the essentials of a good life (e.g. Greek tragedies generally portrayed man's suffering as a consequence of his going beyond the proper measure of things) In this regard, measure was not looked on in its modern sense as being primarily some sort of comparison of an object with an external standard or unit. Rather, this latter procedure was regarded as a kind of outward display or appearance of a deeper 'inner measure', which played an essential role in everything. When something went beyond its proper measure, this meant not merely that it was not conforming to some external standard of what was right but, much more, that it was inwardly out of harmony, so that it was bound to lose its integrity and break up into fragments. One can obtain some insight into this way of thinking by considering the earlier meanings of certain words. Thus, the Latin 'mederi' meaning 'to cure' (the root of the modern 'medicine') is based on a root meaning 'to measure'. This reflects the view that physical health is to be regarded as the outcome of a state of right inward measure in all parts and processes of the body. Similarly, the word 'moderation', which describes one of the prime ancient notions of virtue, is based on the same root, and this shows that such virtue was regarded as the outcome of a right inner measure underlying man's social actions and behaviour. Again, the word 'meditation', which is based on the same root, implies a kind of weighing, pondering, or measuring of the whole process of thought, which could bring the inner activities of the mind to a state of harmonious measure. So, physically, socially and mentally, awareness of the inner measure of things was seen as the essential key to a healthy, happy, harmonious life.

It is clear that measure is to be expressed in more detail through proportion or ratio; and 'ratio' is the Latin word from which our modern 'reason' is derived. In the ancient view, reason is seen as insight into a totality of ratio or proportion, regarded as relevant inwardly to the very nature of things (and not only outwardly as a form of comparison with a standard or unit). Of course, this ratio is not necessarily merely a numerical proportion (though it does, of course, include such proportion). Rather, it is in general a qualitative sort of universal proportion or relationship. Thus, when Newton perceived the insight of universal gravitation, what he saw could be put in this way: 'As the apple falls, so does the moon, and so indeed does everything.' To exhibit the form of the ratio yet more explicitly, one can write:

$$A : B :: C : D :: E : F$$

where A and B represent successive positions of the apple at successive moments of time, C and D those of the moon, and E and F those of any other object.

Whenever we find a theoretical reason for something, we are exemplifying this notion of ratio, in the sense of implying that as the various aspects are related in our idea, so they are related in the thing that the idea is about. The essential reason or ratio of a thing is then the totality of inner proportions in its structure, and in the process in which it forms, maintains itself, and ultimately dissolves. In this view, to understand such ratio is to understand the 'innermost being' of that thing.

It is thus implied that measure is a form of insight into the essence of everything, and that man's perception, following on ways indicated by such insight, will be clear and will thus bring about generally orderly action and harmonious living. In this connection, it is useful to call to mind Ancient Greek notions of measure in music and in the visual arts. These notions emphasized that a grasp of measure was a key to the understanding of harmony in music (e.g., measure as rhythm, right proportion in intensity of sound, right proportion in tonality, etc.). Likewise, in the visual arts, right measure was seen as essential to overall harmony and beauty (e.g., consider the 'Golden Mean'). All of this indicates how far the notion of measure went beyond that of comparison with an external standard, to point to a universal sort of inner ratio or proportion, perceived both through the senses and through the mind.

Of course, as time went on, this notion of measure gradually began to change, to lose its subtlety and to become relatively gross and mechanical. Probably this was because man's notion of measure became more and more routinized and habitual, both with regard to its outward display in measurements relative to an external unit and to its inner significance as universal ratio relevant to physical health, social order, and mental harmony. Men began to learn such notions of measure mechanically, by conforming to the teachings of their elders or their masters, and not creatively through an inner feeling and understanding of the deeper meaning of the ratio or proportion which they were learning. So measure gradually came to be taught as a sort of rule that was to be imposed from outside on the human being, who in turn imposed the corresponding measure physically, socially and mentally, in every context in which he was working. As a result, the prevailing notions of measure were no longer seen as forms of insight. Rather, they appeared to be 'absolute truths about reality as it is', which men seemed always to have known, and whose origin was often explained mythologically as binding injunctions of the Gods, which it would be both dangerous and wicked to question. Thought about measure thus tended to fall mainly into the domain of unconscious habit and, as a result, the forms induced in perception by this thought were now seen as directly observed objective realities, which were essentially independent of how they were thought about.

Even by the time of the Ancient Greeks, this process had gone a long way and, as men realized this, they began to question the notion of measure. Thus Protagoras said: 'Man is the measure of all things', thus emphasizing that measure is not a reality external to man, existing independently of him. But many who were in the habit of looking at everything externally also applied this way of looking to what Protagoras said. Thus, they concluded that measure was something arbitrary, and subject to the capricious choice or taste of each individual. In this way they of course overlooked the fact that measure is a form of insight that has to fit the overall reality in which man lives, as demonstrated by the clarity of perception and harmony of action to which it leads. Such insight can arise properly only when a man works with seriousness and honesty, putting truth and factuality first, rather than his own whims or desires.

The general rigidification and objectification of the notion of measure continued to develop until, in modern times, the very word 'measure' has come to denote mainly a process of comparison of something with an external standard. While the original meaning still survives in some contexts (e.g., art and mathematics) it is generally felt as having only a secondary sort of significance.

Now, in the East the notion of measure has not played nearly so fundamental a role. Rather, in the prevailing philosophy in the Orient, the immeasurable (i.e. that which cannot be named, described, or understood through any form of reason) is regarded as the primary reality. Thus, in Sanskrit (which has an origin common to the Indo-European language group) there is a word 'matra' meaning 'measure', in the musical sense, which is evidently close to the Greek 'metron'. But then there is another word 'maya' obtained from the same root, which means 'illusion'. This is an extraordinarily significant point. Whereas to Western society, as it derives from the Greeks, measure, with all that this word implies, is the very essence of reality, or at least the key to this essence, in the East measure has now come to be regarded commonly as being in some way false and deceitful. In this view the entire structure and order of forms, proportions, and 'ratios' that present themselves to ordinary perception and reason are regarded as a sort of veil, covering the true reality, which cannot be perceived by the senses and of which nothing can be said or thought.

It is clear that the different ways the two societies have developed fit in with their different attitudes to measure. Thus, in the West, society has mainly emphasized the development of science and technology (dependent on measure) while in the East, the main emphasis has gone to religion and philosophy (which are directed ultimately toward the immeasurable).

If one considers this question carefully, one can see that in a certain sense the East was right to see the immeasurable as the primary reality. For, as has already been indicated, measure is an insight created by man. A reality that is beyond man and prior to him cannot depend on such insight. Indeed, the attempt to suppose that measure exists prior to man and independently of him leads, as has been seen, to the 'objectification' of man's insight, so that it becomes rigidified and unable to change, eventually bringing about fragmentation and general confusion in the way described in this chapter.

One may speculate that perhaps in ancient times, the men who were wise enough to see that the immeasurable is the primary reality were also wise enough to see that measure is insight into a secondary and dependent but nonetheless necessary aspect of reality. Thus they may have agreed with the Greeks that insight into measure is capable of helping to bring about order and harmony in our lives, while at the same time, seeing perhaps more deeply, that it cannot be what is most fundamental in this regard.

What they may further have said is that when measure is identified with the very essence of reality, *this* is illusion. But then, when men learned this by conforming to the teachings of tradition, the meaning became largely habitual and mechanical. In the way indicated earlier, the subtlety was lost and men began to say simply: 'measure is illusion'. Thus, both in the East and in the West, true insight may have been turned into something false and misleading by the procedure of learning mechanically through conformity to existent teachings, rather than through a creative and original grasp of the insights implicit in such teachings.

It is of course impossible to go back to a state of wholeness that may have been present before the split between East and West developed (if only because we know little, if anything, about this state). Rather, what is needed is to learn afresh, to observe, and to discover for ourselves the meaning of wholeness. Of course, we have to be cognisant of the teachings of the past, both Western and Eastern, but to imitate these teachings or to try to conform to them would have little value. For, as has been pointed out in this chapter, to develop new insight into fragmentation and wholeness requires a creative work even more difficult than that needed to make fundamental new discoveries in science, or great and original works of art. It might in this context be said that one who is similar to Einstein in creativity is not the one who imitates Einstein's ideas, nor even the one who applies these ideas in new ways, rather, it is the one who learns from Einstein and then goes on to do something original, which is able to assimilate what is valid in Einstein's work and yet goes beyond this work in qualitatively new ways. So what we have to do with regard to the great wisdom from the whole of the past, both in the East and in the West, is to assimilate it and to go on to new and original perception relevant to our present condition of life.

In doing this, it is important that we be clear on the role of techniques, such as those used in various forms of meditation. In a way, techniques of meditation can be looked on as measures (actions ordered by knowledge and reason) which are taken by man to try to reach the immeasurable, i.e., a state of mind in which he ceases to sense a separation between himself and the whole of reality. But clearly, there is a contradiction in such a notion, for the immeasurable is, if anything, just that which cannot be brought within limits determined by man's knowledge and reason.

To be sure, in certain specifiable contexts, technical measures, understood in a right spirit, can lead us to do things from which we can derive insight if we are observant. Such possibilities, however, are limited. Thus, it would be a contradiction in terms to think of formulating techniques for making fundamental new discoveries in science or original and creative works of art, for the very essence of such action is a certain freedom from dependence on others, who would be needed as guides. How can this freedom be transmitted in an activity in which conformity to someone else's knowledge is the main source of energy? And if techniques cannot teach originality and creativity in art and science, how much less is it possible for them to enable us to 'discover the immeasurable'?

Actually, there are no direct and positive things that man can do to get in touch with the immeasurable, for this must be immensely beyond anything that man can grasp with his mind or accomplish with his hands or his instruments. What man *can* do is to give his full attention and creative energies to bring clarity and order into the totality of the field of measure. This involves, of course, not only the outward display of measure in terms of external units but also inward measure, as health of the body, moderation in action, and meditation, which gives insight into the measure of thought. This latter is particularly important because, as has been seen, the illusion that the self and the world are broken into fragments originates in the kind of thought that goes beyond its proper measure and confuses its own product with the same independent reality. To end this illusion requires insight, not only into the world as a whole, but also into how the instrument of thought is working. Such insight implies an original and creative act of perception into all aspects of life, mental and physical, both through the senses and through the mind, and this is perhaps the true meaning of meditation.

As has been seen, fragmentation originates in essence in the fixing of the insights forming our overall self-world view, which follows on our generally mechanical, routinized and habitual modes of thought about these matters. Because the primary reality goes beyond anything that can be contained in such fixed forms of measure, these insights must eventually cease to be adequate, and will thus give rise to various forms of unclarity or confusion. However, when the whole field of measure is open to original and creative insight, without any fixed limits or barriers, then our overall world views will cease to be rigid, and the whole field of measure will come into harmony, as fragmentation within it comes to an end. But original and creative insight within the whole field of measure *is* the action of the immeasurable. For when such insight occurs, the source cannot be within ideas already contained in the field of measure but rather has to be in the immeasurable, which contains the essential formative cause of all that happens in the field of measure. The measurable and the immeasurable are then in harmony and indeed one sees that they are but different ways of considering the one and undivided whole.

When such harmony prevails, man can then not only have insight into the meaning of wholeness but, what is much more significant, he can realize the truth of this insight in every phase and aspect of his life.

As Krishnamurti[1] has brought out with great force and clarity, this requires that man gives his full creative energies. This requires, however, that man gives his full creative energies to the inquiry into the whole field of measure. To do this may perhaps be extremely difficult and arduous, but since everything turns on this, it is surely worthy of the serious attention and utmost consideration of each of us.

2

The rheomode – an experiment with language and thought

1 INTRODUCTION

In the previous chapter it has been pointed out that our thought is fragmented, mainly by our taking it for an image or model of 'what the world is'. The divisions in thought are thus given disproportionate importance, as if they were a widespread and pervasive structure of independently existent actual breaks in 'what is', rather than merely convenient features of description and analysis. Such thought was shown to bring about a thoroughgoing confusion that tends to permeate every phase of life, and that ultimately makes impossible the solution of individual and social problems. We saw the urgent need to end this confusion, through giving careful attention to the one-ness of the content of thought and the actual process of thinking which produces this content.

In this chapter the main emphasis will be to inquire into the role of language structure in helping to bring about this sort of fragmentation in thought. Though language is only *one* of the important factors involved in this tendency, it is clearly of key importance in thought, in communication, and in the organization of human society in general.

Of course, it is possible merely to observe language as it is, and has been, in various differing social groups and periods of history, but what we wish to do in this chapter is to *experiment* with changes in the structure of the common language. In this experimentation our aim is not to produce a well-defined alternative to present language structures. Rather, it is to see what happens to the language function as we change it, and thus perhaps to make possible a certain insight into how language contributes to the general fragmentation. Indeed, one of the best ways of learning how one is conditioned by a habit (such as the common usage of language is, to a large extent) is to give careful and sustained attention to one's overall reaction when one 'makes the test' of seeing what takes place when one is doing something significantly different from the automatic and accustomed function. So, the main point of the work discussed in this chapter is to take a step in what might be an unending experimentation with language (and with thought). That is, we are suggesting that such experimentation is to be considered as a normal activity of the individual and of society (as it has in fact come to be considered over the past few centuries with regard to experimentation with nature and with man himself). Thus, language (along with the thought involved in it) will be seen as a particular field of function among all the rest, so that it ceases to be, in effect, the one field that is exempted from experimental inquiry.

2 AN INQUIRY INTO OUR LANGUAGE

In scientific inquiries a crucial step is to ask the right question. Indeed, each question contains presuppositions, largely implicit. If these presuppositions are wrong or confused, then the question itself is wrong, in the sense that to try to answer it has no meaning. One has thus *to inquire into the appropriateness of the question*. In fact, truly original discoveries in science and in other fields have generally involved such inquiry into old questions, leading to a perception of their inappropriateness, and in this way allowing for the putting forth of new questions. To do this is often very difficult, as these presuppositions tend to be hidden very deep in the structure of our thought. (For example, Einstein saw that questions having to do with space and time and the particle nature of matter, as commonly accepted in the physics of his day, involved confused presuppositions that had to be dropped, and thus he was able to come to ask new questions leading to radically different notions on the subject.)

What, then, will be our question, as we engage in this inquiry into our language (and thought)? We begin with the fact of general fragmentation. We can ask in a preliminary way whether there are any features of the commonly used language which tend to sustain and propagate this fragmentation, as well as, perhaps, to reflect it. A cursory examination shows that a very important feature of this kind is the subject-verb-object structure of sentences, which is common to the grammar and syntax of modern languages. This structure implies that all action arises in a separate entity, the subject, and that, in cases described by a transitive verb, this action crosses over the space between them to another separate entity, the object. (If the verb is intransitive, as in 'he moves', the subject is still considered to be a separate entity but the activity is considered to be either a property of the subject or a reflexive action of the subject, e.g., in the sense that 'he moves' may be taken to mean 'he moves himself'.)

This is a pervasive structure, leading in the whole of life to a function of thought tending to divide things into separate entities, such entities being conceived of as essentially fixed and static in their nature. When this view is carried to its limit, one arrives at the prevailing scientific world view, in which everything is regarded as ultimately constituted out of a set of basic particles of fixed nature.

The subject-verb-object structure of language, along with its world view, tends to impose itself very strongly in our speech, even in those cases in which some attention would reveal its evident inappropriateness. For example, consider the sentence 'It is raining.' Where is the 'It' that would, according to the sentence, be 'the rainer that is doing the raining'? Clearly, it is more accurate to say: 'Rain is going on.' Similarly, we customarily say, 'One elementary particle acts on another', but, as indicated in the previous chapter, each particle is only an abstraction of a relatively invariant form of movement in the whole field of the universe. So it would be more appropriate to say, 'Elementary particles are on-going movements that are mutually dependent because ultimately they merge and interpenetrate.' However, the same sort of description holds also on the larger-scale level. Thus, instead of saying, 'An observer looks at an object', we can more appropriately say, 'Observation is going on, in an undivided movement involving those abstractions customarily called "the human being" and "the object he is looking at".'

These considerations on the overall implications of sentence structures suggest another question. Is it not possible for the syntax and grammatical form of language to be changed so as to give a basic role to the verb rather than to the noun? This would help to end the sort of fragmentation indicated above, for the verb describes actions and movements, which flow into each other and merge, without sharp separations or breaks. Moreover, since movements are in general always themselves changing, they have in them no permanent pattern of fixed form with which separately existent things could be identified. Such an approach to language evidently fits in with the overall world view discussed in the previous chapter, in which movement is, in effect, taken as a primary notion, while apparently static and separately existent things are seen as relatively invariant states of continuing movement (e.g., recall the example of the vortex).

Now, in some ancient languages – for example, Hebrew – the verb was in fact taken as primary, in the sense described above. Thus, the root of almost all words in Hebrew was a certain verbal form, while adverbs, adjectives and nouns were obtained by modifying the verbal form with prefixes, suffixes, and in other ways. However, in modern Hebrew the actual usage is similar to that of English, in that the noun is in fact given a primary role in its meaning even though in the formal grammar all is still built from the verb as a root.

We have to try here, of course, to work with a structure in which the verb has a primary function, and to take this requirement seriously. That is to say, there is no point in using the verb in a formally primary role and to think in terms in which a set of separate and identifiable objects is taken to be what is basic. To say one thing and do another in this way is a form of confusion that would evidently simply add to the general fragmentation rather than help bring it to an end.

Suddenly to invent a whole new language implying a radically different structure of thought is, however, clearly not practicable. What can be done is provisionally and experimentally to introduce a *new mode* of language. Thus, we already have, for example, different moods of the verb, such as the indicative, the subjunctive, the imperative, and we develop skill in the use of language so that each of these moods functions, when it is required, without the need for conscious choice. Similarly, we will now consider a mode in which movement is to be taken as primary in our thinking and in which this notion will be incorporated into the language structure by allowing the verb rather than the noun to play a primary role. As one develops such a mode and works with it for a while, one may obtain the necessary skill in using it, so that it will also come to function whenever it is required, without the need for conscious choice.

For the sake of convenience we shall give this mode a name,

i.e. the *rheomode* ('rheo' is from a Greek verb, meaning 'to flow'). At least in the first instance the rheomode will be an experiment in the use of language, concerned mainly with trying to find out whether it is possible to create a new structure that is not so prone toward fragmentation as is the present one. Evidently, then, our inquiry will have to begin by emphasizing the role of language in shaping our overall world views as well as in expressing them more precisely in the form of general philosophical ideas. For as suggested in the previous chapter these world views and their general expressions (which contain tacit conclusions about everything, including nature, society, ourselves, our language, etc.) are now playing a key role in helping to originate and sustain fragmentation in every aspect of life. So we will start by using the rheomode mainly in an experimental way. As already pointed out, to do this implies giving a kind of careful attention to how thought and language actually work, which goes beyond a mere consideration of their content.

At least in the present inquiry the rheomode will be concerned mainly with questions having to do with the broad and deep implications of our overall world views which now tend to be raised largely in the study of philosophy, psychology, art, science and mathematics, but especially in the study of thought and language themselves. Of course, this sort of question can also be discussed in terms of our present language structure. While this structure is indeed dominated by the divisive form of subject-verb-object, it nevertheless contains a rich and complex variety of other forms, which are used largely *tacitly* and by *implication* (especially in poetry but more generally in all artistic modes of expression). However, the dominant form of subject-verb-object tends continually to lead to fragmentation; and it is evident that the attempt to avoid this fragmentation by skilful use of other features of the language can work only in a limited way, for, by force of habit, we tend sooner or later, especially in broad questions concerning our overall world views, to fall unwittingly into the fragmentary mode of functioning implied by the basic structure. The reason for this is not only that the subject-verb-object form of the language is continually implying an inappropriate division between things but, even more, that the ordinary mode of language tends very strongly to take its own function for granted, and thus it leads us to concentrate almost exclusively on the content under discussion, so that little or no attention is left for the actual symbolic function of the language itself. As pointed out earlier, however, it is here that the primary tendency toward fragmentation originates. For because the ordinary mode of thought and language does not properly call attention to its own function, this latter seems to arise in a reality independent of thought and language, so that the divisions implied in the language structure are then projected, as if they were fragments, corresponding to actual breaks in 'what is'.

Such fragmentary perception may, however, give rise to the illusory impression that adequate attention is indeed already being given to the function of thought and language, and thus may lead to the false conclusion that there is in reality no serious difficulty of the sort described above. One may suppose, for example, that as the function of the world of nature is studied in physics, and that of society is studied in sociology, and that of the mind in psychology, so the function of language is given attention in linguistics. But of course such a notion would be appropriate only if all these fields were actually clearly separated and either constant or slowly changing in their natures, so that the results obtained in each field of specialization would be relevant in all situations and on all occasions in which they might be applied. What we have been emphasizing, however, is that on questions of such broad and deep scope, this sort of separation is not appropriate and that in any case the crucial point is to give attention to the very language (and thought) that is being used, from moment to moment, in the inquiry into the function of language itself, as well as in any other form of inquiry in which one may engage. So it will not be adequate to isolate language as a particular field of inquiry and to regard it as a relatively static thing which changes only slowly (or not at all) as one goes into it.

It is clear, then, that in developing the rheomode, we will have to be especially aware of the need for language properly to call attention to its own function at the very moment in which this is taking place. In this way, we may not only be able to think more coherently about broad questions concerning our general world views, but we may also understand better how the ordinary mode of language functions, so that we may be able to use even this ordinary mode more coherently.

3 THE FORM OF THE RHEOMODE

We now go on to inquire in more detail into what may be a suitable form of expression for the rheomode.

As a first step in this inquiry, we may ask whether the rich and complex informal structure of the commonly used language does not contain, even if perhaps only in a rudimentary or germinal form, some feature that can satisify the need, indicated above, to call attention to the real function of thought and language. If one looks into this question, one can see that there are such features. Indeed, in modern times, the most striking example is the use (and over-use) of the word 'relevant' (which may perhaps be understood as a kind of 'groping' for the attention-calling function that people almost unconsciously feel to be important).

The word 'relevant' derives from a verb 'to relevate', which has dropped out of common usage, whose meaning is 'to lift' (as in 'elevate'). In essence, 'to relevate' means 'to lift into attention', so that the content thus lifted stands out 'in relief'. When a content lifted into attention is coherent or fitting with the context of interest, i.e. when it has some bearing on the context of some relationship to it, then one says that this content is *relevant*; and, of course, when it does not fit in this way, it is said to be *irrelevant*.

As an example, we can take the writings of Lewis Carroll, which are full of humour arising from the use of the irrelevant. Thus, in *Through the Looking Glass*, there is a conversation between the Mad Hatter and the March Hare, containing the sentence: 'This watch doesn't run, even though I used the best butter.' Such a sentence lifts into attention the irrelevant notion that the grade of butter has bearing on the running of watches – a notion that evidently does not fit the context of the actual structure of watches.

In making a statement about relevance, one is treating thought and language as realities, on the same level as the context in which they refer. In effect, one is, at the very moment in which the statement is made, looking or giving attention both to this context and to the overall function of thought and language, to see whether or not they fit each other. Thus, to see the relevance or irrelevance of a statement is primarily an act of perception of a very high order similar to that involved in seeing its truth or falsity. In one sense the question of relevance comes before that of truth, because to ask whether a statement is true of false presupposes that it is relevant (so that to try to assert the truth or falsity of an irrelevant statement is a form of confusion), but in a deeper sense the seeing of relevance or irrelevance is evidently an aspect of the perception of truth in its overall meaning.

Clearly, the act of apprehending relevance or irrelevance cannot be reduced to a technique or a method, determined by some set of rules. Rather, this is an *art*, both in the sense of requiring creative perception and in the sense that this perception has to develop further in a kind of skill (as in the work of the artisan).

Thus it is not right, for example, to regard the division between relevance and irrelevance as a form of accumulated knowledge of properties belonging to statements (e.g., by saying that certain statements 'possess' relevance while others do not). Rather, in each case, the statement of relevance or irrelevance is communicating a perception taking place at the moment of expression, and is the individual context indicated in that moment. As the context in question changes, a statement that was initially relevant may thus cease to be so, or vice versa. Moreover, one cannot even say that a given statement is either relevant or irrelevant, and that this covers all the possibilities. Thus, in many cases, the total context may be such that one cannot clearly perceive whether the statement has bearing or not. This means that one has to learn more, and that the issue is, as it were, in a state of flux. So when relevance or irrelevance is communicated, one has to understand that this is not a hard and fast division between opposing categories but, rather, an expression of an ever-changing perception, in which it is possible, for the moment, to see a fit or non-fit between the content lifted into attention and the context to which it refers.

At present, the question of fitting or non-fitting is discussed through a language structure in which nouns are taken as basic (e.g., by saying 'this notion is relevant'). Such a structure does indeed formally imply a hard and fast division between relevance and irrelevance. So the form of the language is continually introducing a tendency toward fragmentation, even in those very features whose function is to call attention to the wholeness of language and the context in which it is being used.

As already stated we are, of course, often able to overcome this tendency toward fragmentation by using language in a freer, more informal, and 'poetic' way, that properly communicates the truly fluid nature of the difference between relevance and irrelevance. Is it not possible, however, to do this more coherently and effectively by discussing the issue of relevance in terms of the rheomode, in which as suggested earlier, hard and fast divisions do not arise formally, because the verb, rather than the noun, is given a primary role?

To answer this question, we first note that the verb 'to relevate', from which the adjective 'relevant' is derived, ultimately comes from the root 'to levate' (whose meaning is, of course, 'to lift'). As a step in developing the rheomode, we then propose that the verb 'to levate' shall mean, 'The spontaneous and unrestricted act of lifting into attention any content whatsoever, which includes the lifting into attention of the question of whether this content fits a broader context or not, as well as that of lifting into attention the very function of *calling attention* which is initiated by the verb itself.' This implies an unrestricted breadth and depth of meaning, that is not fixed within static limits.

We then introduce the verb 'to re-levate'. This means: 'To lift a certain content into attention again, for a particular context, as indicated by thought and language.' Here, it has to be emphasized that 're' signifies 'again', i.e. on another occasion. It evidently implies *time* and similarity (as well as difference, since each occasion is not only similar but also different).

As pointed out earlier, it then requires an act of perception to see, in each case, whether the content thus 'lifted again' fits the observed context or not. In those cases in which this act of perception reveals a fit, we say: 'to re-levate is re-levant' (note that the use of the hyphen is essential here, and that the word should be pronounced with a break, as indicated by the hyphen). Of course, in those cases in which perception reveals non-fitting, we say 'to re-levate is irre-levant'.

We see, then, that adjectives have been built from the verb as a root form. Nouns also can be constructed in this way, and they will signify not separate objects but, rather, *continuing states* of activity of the particular form indicated by the verbs. Thus, the noun 're-levation' means 'a continuing state of lifting a *given* content into attention'.

To go on with re-levation when to do so is irre-levant will, however, be called 'irre-levation'. In essence, irre-levation implies that there is not proper attention. When some content is irrelevant, it should normally sooner or later be dropped. If this does not happen, then one is, in some sense, not watchful or alert. Thus, irre-levation implies the need to give attention to the fact that there is not proper attention. *Attention to such failure of attention is of course the very act that ends irre-levation.*

Finally, we shall introduce the noun form 'levation', which signifies a sort of generalized and unrestricted totality of acts of lifting into attention (note that this differs from the 'to levate', which signifies a single spontaneous and unrestricted act of lifting into attention).

Clearly, the above way of using a structure of language form built from a root verb enables us to discuss what is commonly meant by 'relevance' in a way that is free of fragmentation, for we are no longer being led, by the form of the language, to consider something called relevance as if it were a separate and fixed quality. Even more important, we are not establishing a division between what the verb 'to levate' means and the actual function that takes place when we use this verb. That is to say, 'to levate' is not only to attend to the thought of lifting an unrestricted content into attention but it is also to engage in the very act of lifting such an unrestricted content into attention. The thought is thus not a mere abstraction, with no concrete perception to which it can refer. Rather, something is actually going on which fits the meaning of the word, and one can, at the very moment of using the word, perceive the fit between this meaning and what is going on. So the content of thought and its actual function are seen and felt as one, and thus one understands what it can mean for fragmentation to cease, at its very origin.

Evidently, it is possible to generalize this way of building up language forms so that any verb may be taken as the root form. We shall then say that the rheomode is in essence characterized by this way of using a verb.

As an example, let us consider the Latin verb 'videre', meaning 'to see', which is used in English in such forms as 'video'. We then introduce the root verbal form 'to vidate'. This does not mean merely 'to see' in the visual sense, but we shall take it to refer to every aspect of perception including even the act of understanding, which is the apprehension of a totality, that includes sense perception, intellect, feeling, etc. (e.g., in the common language 'to understand' and 'to see' may be used interchangeably). So the word 'to vidate' will call attention to a spontaneous and unrestricted act of perception of any sort whatsoever, including perception of whether what is seen fits or does not fit 'what is', as well as perception even of the very attention-calling function of the word itself. Thus, as happens with 'to levate', there is no division between the content (meaning) of this word and the total function to which it gives rise.

We then consider the verb 'to re-vidate', which means to perceive a given content *again*, as indicated by a word or thought. If this content is seen to fit the indicated context, then we say: 'to re-vidate is re-vidant'. If it is seen not to fit, then of course we say: 'to re-vidate is irre-vidant' (which means, in ordinary usage, that this was a mistaken or illusory perception).

'Re-vidation' is then a continuing state of perceiving a certain content, while 'irre-vidation' is a continuing state of being caught in illusion or delusion, with regard to a certain content. Evidently (as with irre-levation) irre-vidation implies a failure of attention, and to attend to this failure of attention is to end irre-vidation.

Finally, the noun 'vidation' means an unrestricted and generalized totality of acts of perception. Clearly, *vidation* is not to be sharply distinguished from *levation*. In an act of vidation, it is necessary to levate a content into attention, and in an act of levation, it is necessary to vidate this content. So the two movements of levation and vidation merge and interpenetrate. Each of these words merely emphasizes (i.e., re-levates) a certain aspect of movement in general. It will become evident that this will be true of all verbal roots in the rheomode. They all imply each other, and pass into each other. Thus, the rheomode will reveal a certain wholeness, that is not characteristic of the ordinary use of language (though it is there potentially, in the sense that if we start with movement as primary, then we have likewise to say that all movements shade into each other, to merge and interpenetrate).

Let us now go on to consider the verb 'to divide'. We shall take this to be a combination of the verb 'videre' and the prefix 'di', meaning 'separate'. So, 'to divide' is to be considered[1] as meaning 'to see as separate'.

We thus introduce the verb[2] 'to di-vidate'. This word calls attention to the spontaneous act of seeing things as separate, in any form whatsoever, including the act of seeing whether or not the perception fits 'what is', and even that of seeing how the attention-calling function of this word has a form of inherent division in it. With regard to this last point, we note that merely to consider the word 'di-vidate' makes it clear that this is different from the word 'vidate' from which it has been derived. So, to di-vidate implies not only a *content* (or meaning) of division but also that the very use of this word produces a function for which the notion of division is seen to provide a description that fits.

We now consider the verb 'to re-dividate', which means through thought and language to perceive a given content again in terms of a particular kind of separation or division. If to do this is seen to fit the indicated context, then we say that 'to re-dividate is re-dividant'. If it is seen not to fit, we say that to 're-dividate is irre-dividant'.

Re-dividation is then a continuing state of seeing a certain content in the form of separation or division. Irre-dividation is a continuing state of seeing separation where, in the ordinary language, we would say that separation is irrelevant.

Irre-dividation is clearly essentially the same as fragmentation.

So it becomes evident that fragmentation cannot possibly be a good thing, for it means not merely to see things as separate but to persist in doing this in a context in which this way of seeing does not fit. To go on indefinitely with irre-dividation is possible only through a failure of attention. Thus irre-dividation comes to an end in the very act of giving attention to this failure of attention.

Finally, of course, the noun 'dividation' means an unrestricted and generalized totality of acts of seeing things as separate. As has been indicated earlier, di-vidation implies a division in the attention-calling function of the word, in the sense that di-vidation is seen to be different from vidation. Nevertheless, this difference holds only in some limited context and is not to be taken as a fragmentation, or actual break, between the meanings and functions of the two words. Rather, their very forms indicate that di-vidation is a kind of vidation, indeed a special case of the latter. So ultimately, wholeness is primary, in the sense that these meanings and functions pass into each other to merge and interpenetrate. Division is thus seen to be a convenient means of giving a more articulated and detailed description to this whole, rather than a fragmentation of 'what is'.

The movement from division to one-ness of perception is through the action of *ordering*. (A more detailed discussion of this is given in chapter 5.) For example, a ruler may be divided into inches, but this set of divisions is introduced into our thinking only as a convenient means of expressing a *simple sequential order*, by which we can communicate and understand something that has bearing on some whole object, which is measured with the aid of such a ruler.

This simple notion of a sequential order, expressed in terms of regular divisions in a line on a scale, helps to direct us in our constructional work, our travels and movements on the surface of the Earth and in space, and in a wide range of general practical and scientific activities. But, of course, more complex orders are possible, and these have to be expressed in terms of more subtle divisions and categories of thought, which are significant for more subtle forms of movement. Thus, there is the movement of growth, development and evolution of living beings, the movement of a symphony, the movement that is the essence of life itself, etc. These evidently have to be described in different ways, that cannot generally be reduced to a description in terms of simple sequential orders.

Beyond all these orders is that of the movement of attention.

This movement has to have an order that fits the order in that which is to be observed, or else we will miss seeing what is to be seen. For example, if we try to listen to a symphony while our attention is directed mainly to a sequential time order as indicated by a clock, we will fail to listen to the subtle orders that constitute the essential meaning of the music. Evidently, our ability to perceive and understand is limited by the freedom with which the ordering of attention can change, so as to fit the order that is to be observed.

It is clear, then, that in the understanding of the true meaning of the divisions of thought and language established for our convenience the notion of order plays a key role. To discuss this notion in the rheomode let us then introduce the verbal root form 'to ordinate'. This word calls attention to a spontaneous and unrestricted act of ordering of any sort whatsoever, including the ordering involved in seeing whether any particular order fits or does not fit some observed context, and even the ordering which arises in the attention-calling function itself. So 'to ordinate' does not primarily mean 'to think about an order' but, rather, to engage in the very act of ordering attention, while attention is given also to one's thoughts about order. Once again, we see the wholeness of the meaning of a word and its overall function, which is an essential aspect of the rheomode.

'To re-ordinate' is then to call attention again to a given order, by means of language and thought. If this order is seen to fit that which is to be observed in the context under discussion, we say that 'to re-ordinate is re-ordinant'. If it is seen not to fit, we say that 'to re-ordinate is irre-ordinant' (e.g., as in the application of a linear grid to a complex maze of alleyways).

The noun 're-ordination' then describes a continuing state of calling attention to a certain order. A persistent state of re-ordination in an irre-ordinant context will then be called 'irre-ordination'. As happens with all other verbs, irre-ordination is possible only through a failure of attention, and comes to an end when attention is given to this failure of attention.

Finally, the noun 'ordination' means, of course, an unrestricted and generalized totality of acts of ordering. Evidently, ordination implies levation, vidation and di-vidation, and ultimately, all these latter imply ordination. Thus, to see whether a given content is re-levant, attention has to be suitably ordered to perceive this content; a suitable set of divisions or categories will have to be set up in thought, etc., etc.

Enough has been said of the rheomode at least to indicate in general how it works. At this point it may, however, be useful to display the overall structure of the rheomode by listing the words that have thus far been used:

> Levate, re-levate, re-levant, irre-levant, levation, re-levation, irre-levation.
> Vidate, re-vidate, re-vidant, irre-vidant, vidation, re-vidation, irre-vidation.
> Di-vidate, re-dividate, re-dividant, irre-dividant, di-vidation, re-dividation, irre-dividation.
> Ordinate, re-ordinate, re-ordinant, irre-ordinant, ordination, re-ordination, irre-ordination.

It should be noted that the rheomode involves, in the first instance, a new grammatical construction, in which verbs are used in a new way. However, what is further novel in it is that the syntax extends not only to the arrangement of words that may be regarded as already given, but also to a systematic set of rules for the formation of new words.

Of course, such word formation has always gone on in most languages (e.g. 'relevant' is built from the root 'levate' with the prefix 're' and the suffix 'ate' replaced by 'ant'), but this kind of construction has tended to arise mainly in a fortuitous way, probably as a result of the need to express various useful relationships. In any case, once the words have been put together the prevailing tendency has been to lose sight of the fact that this has happened and to regard each word as an 'elementary unit', so that the origin of such words in a construction is, in effect, treated as having no bearing on its meaning. In the rheomode, however, the word construction is not fortuitous, but plays a primary role in making possible a whole new mode of language, while the activity of word construction is continually being brought to our notice because the meanings depend in an essential way on the forms of such constructions.

It is perhaps useful here to make a kind of comparison with what has happened in the development of science. As seen in chapter 1 the prevailing scientific world view has generally been to suppose that, at bottom, everything is to be described in terms of the results of combinations of certain 'particle' units, considered to be basic. This attitude is evidently in accord with the prevailing tendency in the ordinary mode of language to treat words as 'elementary units' which, one supposes, can be combined to express anything whatsoever that is capable of being said.

New words can, of course, be brought in to enrich discourse in the ordinary mode of language (just as new basic particles can be introduced in physics) but, in the rheomode, one has begun to go further and to treat the construction of words as not essentially different from the construction of phrases, sentences, paragraphs, etc. Thus, the 'atomistic' attitude to words has been dropped and instead our point of view is rather similar to that of field theory in physics, in which 'particles' are only convenient abstractions from the whole movement. Similarly, we may say that language is an undivided field of movement, involving sound, meaning, attention-calling, emotional and muscular reflexes, etc. It is somewhat arbitrary to give the present excessive significance to the breaks between words. Actually, the relationships between parts of a word may, in general, be of much the same sort as those between different words. So the word ceases to be taken as an 'indivisible atom of meaning' and instead it is seen as no more than a convenient marker in the whole movement of language, neither more nor less fundamental than the clause, the sentence, the paragraph, the system of paragraphs, etc. (This means that giving attention in this way to the components of words is not primarily an attitude of analysis but, rather, an approach that allows for the unrestricted flow of meaning.)

Some insight into the meaning of this change of attitude to words is given by considering language as a particular form of order. This is to say, language not only calls attention to order. It *is* an order of sounds, words, structures of words, nuances of phrase and gesture, etc. Evidently, the meaning of a communication through language depends, in an essential way, on the order that language *is*. This order is more like that of a symphony in which each aspect and movement has to be understood in the light of its relationship to the whole, rather than like the simple sequential order of a clock or a ruler; and since (as has been pointed out here) the order of sounds *within* a word is an inseparable aspect of the whole meaning, we can develop rules of grammar and syntax that use this order in a systematic way to enrich and enhance the possibilities of the language for communication and for thinking.

4 TRUTH AND FACT IN THE RHEOMODE

In the ordinary mode of language, truth is taken as a noun, which thus stands for something that can be grasped once and for all or which can at least be approached, step by step. Or else, the possibility of being either true or false may be taken as a *property* of statements. However, as indicated earlier, truth and falsity have actually, like relevance and irrelevance, to be seen from moment to moment, in an act of perception of a very high order. Thus, the truth or falsity *in content* of a statement is apprehended by observing whether or not this content fits a broader context which is indicated either in the statement itself or by some action or gesture (such as pointing) that goes together with the statement. Moreover, when we come to statements about world views, which have to do with 'the totality of all that is', there is no clearly definable context to which they can refer and so we have to emphasize *truth in function*, i.e. the possibility of free movement and change in our general notions of reality as a whole, so as to allow for a continual fitting to new experience, going beyond the limits of fitting of older notions of this kind. (See chapters 3 and 7 for a further discussion of this.)

It is clear, then, that the ordinary mode of language is very unsuitable for discussing questions of truth and falsity, because it tends to treat each truth as a separate fragment, that is essentially fixed and static in its nature. It will thus be interesting to experiment with the use of the rheomode, to see in what way this can allow us to discuss the question of truth more fittingly and coherently.

We shall begin by considering the Latin 'verus', meaning 'true'. So we shall introduce the root verbal form 'to verrate'. (The double 'r' is brought in here to avoid a certain confusion of a kind that will be evident as we proceed.) This word calls attention, in the manner discussed in the previous section, to a spontaneous and unrestricted act of seeing truth in any form whatsoever, including the act of seeing whether this perception fits or does not fit that which is perceived actually to happen in the apprehension of truth, as well as seeing the truth in the attention-calling function of the word itself. So, 'to verrate' is to be in the act of perceiving truth, as well as to be attending to what truth means.

To re-verrate, then, is to call attention again, by means of thought and language, to a particular truth in a given context. If this is seen to fit what is to be observed in this context, we say that *to re-verrate is re-verrant*, and if it is seen not to fit, we say that *to re-verrate is irre-verrant* (i.e. a particular truth ceases to be valid when repeated and extended into a context that is beyond its proper limits).

We see, then, that the question of truth is no longer being discussed in terms of separate and essentially static fragments.

Rather, our attention is called to the general act of *verration*, and to its continuation in a particular context as *re-verration* and *irre-verration*. (Irre-verration, i.e. the persistent holding to a truth beyond its proper limits, has evidently been one of the major sources of illusion and delusion throughout the whole of history and in every phase of life.) Verration is to be seen as a flowing movement, which merges and interpenetrates with levation, vidation, di-vidation, ordination, and indeed with all the other movements that will be indicated in the subsequent development of the rheomode.

Now, when we discuss truth in the ordinary mode, we are inevitably brought to consider what is to be meant by *the fact*. Thus, in some sense, to say: 'This is a fact' implies that the content of the statement in question is true. However, the root meaning of the word 'fact' is 'that which has been made' (e.g., as in 'manufacture'). This meaning does have bearing here because, as is evident, in some sense we actually do 'make' the fact: for this fact depends not only on the context that is being observed and on our immediate perception, it also depends on how our perceptions are shaped by our thoughts, as well as on what we *do*, to test our conclusions, and to apply them in practical activities.

Let us now go on to experiment with the use of the rheomode, to see where this leads when we consider what is meant by 'the fact'. We thus introduce the root verb 'to factate', meaning a spontaneous and unrestricted attention to consciously directed human activity in *making* or *doing any sort of thing whatsover*[3] (and this, of course, includes the 'making' or 'doing' of the attention-calling function of the word itself). To re-factate is, then, through thought and language, to call attention again to such an activity of 'making' or 'doing' in a particular context. If this activity is seen to fit within the context (i.e. if what we are doing 'works') then we say 'to re-factate is re-factant' and if it is seen not to fit, we say 'to re-factuate is irre-factant'.

Clearly, a great deal of what is ordinarily meant by the truth or falsity of a statement is contained in the implication of the words 're-factant' and 'irre-factant'. Thus it is evident that when true notions are applied in practice, they will generally lead to our doing something that 'works', while false notions will lead to activities that 'do not work'.

Of course, we have to be careful here not to identify truth as nothing more than 'that which works' since, as has been seen, truth is a whole movement, going far beyond the limited domain of our consciously directed functional activities. So, although the statement 're-verration is re-factant' is correct as far as it goes, it is important to keep in mind that this calls attention only to a certain aspect of what is to be meant by truth. Indeed, it does not even cover all that is meant by *fact*. Far more is involved in establishing the fact than merely to observe that our knowledge is re-factant, i.e. that it has generally led us successfully to achieve the goals that were originally projected in thought. In addition, the fact has to be *tested* continually, through further observation and experience. The primary aim of such testing is not the production of some desired result or end but, rather, it is to see whether the fact will 'stand up', even when the context to which it refers is observed again and again, either in essentially the same way as before, or in new ways that may have bearing on this context. In science, such testing is carried out through experiments, which not only have to be reproducible but which also have to fit in with 'cross-checks' provided by other experiments that are significant in the context of interest. More generally, experience as a whole is always providing a similar sort of test, provided that we are alert and observant to see what it actually indicates.

When we say 'this is a fact' we then imply a certain ability of the fact to 'stand up to' a wide range of different kinds of testing. Thus, the fact is *established*, i.e. it is shown to be *stable*, in the sense that it is not liable to collapse, or to be nullified at any moment, in a subsequent observation of the general sort that has already been carried out. Of course, this stability is only relative, because the fact is always being tested again and again, both in ways that are familiar and in new ways that are continually being explored. So it may be refined, modified, and even radically changed, through further observation, experiment and experience. But in order to be a 'real fact', it evidently has, in this way, to remain *constantly* valid, at least in certain contexts or over a certain period of time.

To lay the ground for discussing this aspect of the fact in the rheomode, we first note that the word 'constant' is derived from a now obsolete verb 'to constate', which means 'to establish', 'to ascertain', or 'to confirm'. This meaning is made even more evident by considering the Latin root 'constare' ('stare' meaning 'to stand' and 'con' meaning 'together'). Thus, we can say that in the activity of testing, we 'constate' the fact; so that is established and 'stands together firmly', as a coherent body, which is able in a certain relative sense, to 'stand up' to being put to the test. Thus, within certain limits, the fact remains *con-stant*.

Actually, the very closely related word 'constater' is used in modern French, in much the sense that has been indicated above. In a certain way, it covers what is meant here better than 'constatc' because it is derived from the Latin 'constat' which is the past participle of 'constare', and thus its root meaning would be 'to have stood together'. This fits together quite well with 'fact' or 'that which *has* been made'.

To consider these questions in the rheomode, we then introduce the root verb 'to con-statate'. This means 'to give spontaneous and unrestricted attention to how any sort of action or movement whatsoever is established in a relatively constant form that stands together relatively stably, including the action of establishing a body of fact that stands together in this way, and even the action of this very word in helping to establish the fact about the function of language itself'.

To re-constatate is then by means of word and thought, to call attention again to a particular action or movement of this kind in a given context. If this latter is seen to fit within the context in question, we say: 'to re-constate is re-constatant', and if it is seen not to fit, we say: 'to re-constate is irre-constatant' (e.g. the fact as it had previously been established is not found factually to 'stand up' to further observation and experience).

The noun form 're-constation' then signifies a particular kind of continuing *state* of action or movement in a given context, that 'stands together' in a relatively constant way, whether this be our own action in establishing a fact, or any other kind of movement that can be described as established or stable in form. It may thus, in the first instance, refer to the possibility of confirming again and again, in a series of acts of observation or experimentation, that 'the fact still stands'; or it may refer to a certain continuing state of movement (or of affairs) which 'still stands' in an overall reality including and going beyond our acts of observation and experimentation. Finally it may refer to the verbal activity of making a statement (i.e. *state*-ment) by which what one person re-constatates can be communicated, to be re-constatated by other people. That is to say, a re-constatation is, in ordinary use of language, 'an established fact' or 'the actual state of movement or of affairs that the fact is about' or 'the verbal statement of the fact'. So we do not make a sharp distinction between the act of perception and experimentation, the action of that which we perceive and of which we experiment, and the activity of communicating verbally about what we have observed and done. All of these are regarded as sides or aspects of an unbroken and undivided whole movement, which are closely related, both in function and in content (and thus we do not fall into a fragmentary division between our 'inward' mental activities and their 'outward' function).

Evidently, this use of the rheomode fits very well with the world view in which apparently static things are likewise seen as abstractions of relatively invariant aspects from an unbroken and undivided whole movement. However, it goes further in implying that the fact about such things is itself abstracted as just that relatively constant aspect of the whole movement appearing in perception and experienced in action, which 'stands together' in a continuing state, and which is thus suitable for communication in the form of a statement.

5 THE RHEOMODE AND ITS IMPLICATIONS FOR OUR OVERALL WORLD VIEW

In seeing (as pointed out in the previous section) that the rheomode does not allow us to discuss the observed fact in terms of separately existent things of an essentially static nature, we are led to note that the use of the rheomode has implications for our general world view. Indeed, as has already been brought out to some extent, every language form carries a kind of dominant or prevailing world view, which tends to function in our thinking and in our perception whenever it is used, so that to give a clear expression of a world view contrary to the one implied in the primary structure of a language is usually very difficult. It is therefore necessary in the study of any general language form to give serious and sustained attention to its world view, both in content and in function.

As indicated earlier, one of the major defects of the ordinary mode of using language is just its general implication that it is not restricting the world view in any way at all, and that in any case questions of world view have to do only with 'one's own particular philosophy', rather than with the content and function of our language, or with the way in which we tend to experience the overall reality in which we live. By thus making us believe that our world view is only a relatively unimportant matter, perhaps involving mainly one's personal taste or choice, the ordinary mode of language leads us to fail to give attention to the actual function of the divisive world view that pervades this mode, so that the automatic and habitual operation of our thought and language is then able to project these divisions (in the manner discussed earlier) as if they were actual fragmentary breaks in the nature of 'what is'. It is thus essential to be aware of the world view implied in each form of language, and to be watchful and alert, to be ready to see when this world view ceases to fit actual observation and experience, as these are extended beyond certain limits.

It has become evident in this chapter that the world view implied in the rheomode is in essence that described in the first chapter, which is expressed by saying that *all* is an unbroken and undivided whole movement, and that each 'thing' is abstracted only as a relatively invariant side or aspect of this movement. It is clear, therefore, that the rheomode implies a world view quite different from that of the usual language structure. More specifically, we see that the mere act of seriously considering such a new mode of language and observing how it works can help draw our attention to the way in which our ordinary language structure puts strong and subtle pressures on us to hold to a fragmentary world view. Whether it would be useful to go further, however, and to try to introduce the rheomode into active usage, it is not possible to say at present, though perhaps some such development may eventually be found to be helpful.

3

Reality and knowledge considered as process

1 INTRODUCTION

The notion that reality is to be understood as process is an ancient one, going back at least to Heraclitus, who said that everything flows. In more modern times, Whitehead[1] was the first to give this notion a systematic and extensive development. In this chapter I shall discuss the question of the relationship between reality and knowledge from such a point of view. However, while my explicit starting point is generally similar to that of Whitehead, some implications will emerge that may be significantly different from those of his work.

I regard the essence of the notion of process as given by the statement: Not only is everything changing, but all *is* flux. That is to say, *what is* the process of becoming itself, while all objects, events, entities, conditions, structures, etc., are forms that can be abstracted from this process.

The best image of process is perhaps that of the flowing stream, whose substance is never the same. On this stream, one may see an ever-changing pattern of vortices, ripples, waves, splashes, etc., which evidently have no independent existence as such. Rather, they are abstracted from the flowing movement, arising and vanishing in the total process of the flow. Such transitory subsistence as may be possessed by these abstracted forms implies only a relative independence or autonomy of behaviour, rather than absolutely independent existence as ultimate substances. (See chapter 1 for a further discussion of this notion.)

Of course, modern physics states that actual streams (e.g., of water) are composed of atoms, which are in turn composed of 'elementary particles', such as electrons, protons, neutrons, etc. For a long time it was thought that these latter are the 'ultimate substance' of the whole of reality, and that all flowing movements, such as those of streams, must reduce to forms abstracted from the motions through space of collections of interacting particles. However, it has been found that even the 'elementary particles' can be created, annihilated and transformed, and this indicates that not even these can be ultimate substances but, rather, that they too are relatively constant forms, abstracted from some deeper level of movement.

One may suppose that this deeper level of movement may be analysable into yet finer particles which will perhaps turn out to be the ultimate substance of the whole of reality. However, the notion that all is flux, into which we are inquiring here, denies such a supposition. Rather, it implies that any describable event, object, entity, etc., is an abstraction from an unknown and undefinable totality of flowing movement. This means that no matter how far our knowledge of the laws of physics may go, the content of these laws will still deal with such abstractions, having only a relative independence of existence and independence of behaviour. So one will not be led to suppose that *all* properties of collections of objects, events, etc., will have to be explainable in terms of some knowable set of ultimate substances. At any stage, further properties of such collections may arise, whose ultimate ground is to be regarded as the unknown totality of the universal flux.

Having discussed what the notion of process implies concerning the nature of reality, let us now consider how this notion should bear on the nature of knowledge. Clearly, to be consistent, one has to say that knowledge, too, is a process, an abstraction from the one total flux, which latter is therefore the ground both of reality and of knowledge of this reality. Of course, one may fairly readily verbalize such a notion, but in actual fact it is very difficult not to fall into the almost universal tendency to treat our knowledge as a set of basically fixed truths, and thus not of the nature of process (e.g., one may admit that knowledge is always changing but say that it is accumulative, thus implying that its basic elements are permanent truths which we have to discover). Indeed, even to assert any absolutely invariant element of knowledge (such as 'all is flux') is to establish in the field of knowledge something that is permanent; but if *all* is flux, then every part of knowledge must have its being as an abstracted form in the process of becoming, so that there can be no absolutely invariant elements of knowledge.

Is it possible to be free of this contradiction, in the sense that one could understand not only reality, but also *all* knowledge, as grounded in the flowing movement? Or must one necessarily regard *some* elements of knowledge (e.g., those concerning the nature of process) as absolute truths, beyond the flux of process? It is to this question that we shall address ourselves in this chapter.

2 THOUGHT AND INTELLIGENCE

To inquire into the question of how knowledge is to be understood as process, we first note that all knowledge is produced, displayed, communicated, transformed, and applied in *thought*. Thought, considered in its *movement of becoming* (and not merely in its content of relatively well-defined images and ideas) *is* indeed the process in which knowledge has its actual and concrete existence. (This has been discussed in the Introduction.)

What is the process of thought? Thought is, in essence, the active response of memory in every phase of life. We include in thought the intellectual, emotional, sensuous, muscular and physical responses of memory. These are all aspects of one indissoluble process. To treat them separately makes for fragmentation and confusion. All these are one process of response of memory to each actual situation, which response in turn leads to a further contribution to memory, thus conditioning the next thought.

One of the earliest and most primitive forms of thought is, for example, just the memory of pleasure or pain, in conjunction with a visual, auditory, or olfactory image that may be evoked by an object or a situation. It is common in our culture to regard memories involving image content as separate from those involving feeling. It is clear, however, that the *whole meaning* of such a memory is just the conjunction of the image with its feeling, which (along with the intellectual content and the physical reaction) constitutes the totality of the judgment as to whether what is remembered is good or bad, desirable or not, etc.

It is clear that thought, considered in this way as the response of memory, is basically mechanical in its order of operation. Either it is a repetition of some previously existent structure drawn from memory, or else it is some combination arrangement and organization of these memories into further structures of ideas and concepts, categories, etc. These combinations may possess a certain kind of novelty resulting from the fortuitous interplay of elements of memory, but it is clear that such novelty is still essentially mechanical (like the new combinations appearing in a kaleidoscope).

There is in this mechanical process no inherent reason why the thoughts that arise should be relevant or fitting to the actual situation that evokes them. The perception of whether or not any particular thoughts are relevant or fitting requires the operation of an energy that is not mechanical, an energy that we shall call *intelligence*. This latter is able to perceive a new order or a new structure, that is not just a modification of what is already known or present in memory. For example, one may be working on a puzzling problem for a long time. Suddenly, in a flash of understanding, one may see the irrelevance of one's whole way of thinking about the problem, along with a different approach in which all the elements fit in a new order and in a new structure. Clearly, such a flash is essentially an *act of perception*, rather than a process of thought (a similar notion was discussed in chapter 1), though later it may be expressed in thought. What is involved in this act is *perception through the mind*, of abstract orders and relationships such as indentity and difference, separation and connection, necessity and contingency, cause and effect, etc.

We have thus put together all the basically mechanical and conditioned responses of memory under one word or symbol, i.e. thought, and we have distinguished this from the fresh, original and unconditioned response of intelligence (or intelligent perception) in which something new may arise. At this point, however, one may ask: 'How can one know that such an unconditioned response is at all possible?' This is a vast question, which cannot be discussed fully here. However, it can be pointed out here that at least implicitly everybody does in fact accept the notion that intelligence is not conditioned (and, indeed, that one cannot consistently do otherwise).

Consider, for example, an attempt to assert that all of man's actions are conditioned and mechanical. Typically, such a view has taken one of two forms: Either it is said that man is basically a product of his hereditary constitution, or else that he is determined entirely by environmental factors. However, one could ask of the man who believed in hereditary determination whether his own statement asserting this belief was nothing but the product of his heredity. In other words, is he compelled by his genetic structure to make such an utterance? Similarly, one may ask of the man who believes in environmental determination, whether the assertion of such a belief is nothing but the spouting forth of words in patterns to which he was conditioned by his environment. Evidently, in both cases (as well as in the case of one who asserted that man is completely conditioned by heredity *plus* environment) the answer would have to be in the negative, for otherwise the speakers would be denying the very possibility that what they said could have meaning. Indeed, it is necessarily implied, in any statement, that the speaker is capable of talking from intelligent perception, which is in turn capable of a *truth* that is not merely the result of a mechanism based on meaning or skills acquired in the past. So we see that no one can avoid implying, by his mode of communication, that he accepts at least the possibility of that free, unconditioned perception that we have called intelligence.

Now, there is a great deal of evidence indicating that thought is basically a material process. For example, it has been observed in a wide variety of contexts that thought is inseparable from electrical and chemical activity in the brain and nervous system, and from concomitant tensions and movements of muscles. Would one then say that intelligence is a similar process, though perhaps of a more subtle nature?

It is implied in the view we are suggesting here that this is not so. If intelligence is to be an unconditioned act of perception, its ground cannot be in structures such as cells, molecules, atoms, elementary particles, etc. Ultimately, anything that is determined by the laws of such structures must be in the field of what can be known, i.e. stored up in memory, and thus it will have to have the mechanical nature of anything that can be assimilated in the basically mechanical character of the process of thought. The actual operation of intelligence is thus beyond the possibility of being determined or conditioned by factors that can be included in any knowable law. So, we see that the ground of intelligence must be in the undetermined and unknown flux, that is also the ground of all definable forms of matter. Intelligence is thus not deducible or explainable on the basis of any branch of knowledge (e.g., physics or biology). Its origin is deeper and more inward than any knowable order that could describe it. (Indeed, it has to comprehend the very order of definable forms of matter through which we would hope to comprehend intelligence.)

What, then, is the relationship of intelligence to thought? Briefly, one can say that when thought functions on its own, it is mechanical and not intelligent, because it imposes its own generally irrelevant and unsuitable order drawn from memory. Thought is, however, capable of responding, not only from memory but also to the unconditioned perception of intelligence that can see, in each case, whether or not a particular line of thought is relevant and fitting.

One may perhaps usefully consider here the image of a radio receiver. When the output of the receiver 'feeds back' into the input, the receiver operates on its own, to produce mainly irrelevant and meaningless noise, but when it is sensitive to the signal on the radio wave, its own order of inner movement of electric currents (transformed into sound waves) is parallel to the order in the signal and thus the receiver serves to bring a meaningful order originating beyond the level of its own structure into movements on the level of its own structure. One might then suggest that in intelligent perception, the brain and nervous system respond directly to an order in the universal and unknown flux that cannot be reduced to anything that could be defined in terms of knowable structures.

Intelligence and material process have thus a single origin, which is ultimately the unknown totality of the universal flux. In a certain sense, this implies that what have been commonly called mind and matter are abstractions from the universal flux, and that both are to be regarded as different and relatively autonomous orders within the one whole movement. (This notion is discussed further in chapter 7.) It is thought responding to intelligent perception which is capable of bringing about an overall harmony or fitting between mind and matter.

3 THE THING AND THE THOUGHT

Given that thought is a material process that may be relevant in some more general context when it moves in parallel with intelligent perception, one is now led to inquire into the relationship between thought and reality. Thus, it is commonly believed that the content of thought is in some kind of reflective correspondence with 'real things', perhaps being a kind of copy, or image, or imitation of things, perhaps a kind of 'map' of things, or perhaps (along lines similar to those suggested by Plato) a grasp of the essential and innermost forms of things.

Are any of these views correct? Or is the question itself not in need of further clarification? For it presupposes that we know what is meant by the 'real thing' and by the distinction between reality and thought. But this is just what is not properly understood (e.g., even the relatively sophisticated Kantian notion of 'thing in itself' is just as unclear as the naïve idea of 'real thing').

We may perhaps obtain a clue here by going into the origins of words such as 'thing' and 'reality'. The study of origins of words may be regarded as a sort of archaeology of our thought process, in the sense that the traces of earlier forms of thought can be found by observations made in this field. As in the study of human society, clues coming from archaeological inquiries can often help us to understand the present situation better.

Now the word 'thing' goes back to various old English words[2] whose significance includes 'object', 'action', 'event', 'condition', 'meeting', and is related to words meaning 'to determine', 'to settle', and, perhaps, to 'time' or 'season'. The original meaning might thus have been 'something occurring at a given time, or under certain conditions'. (Compare to the German 'bedingen', meaning 'to make conditions', or 'to determine', which could perhaps be rendered into English as 'to bething'.) All these meanings indicate that the word 'thing' arose as a highly generalized indication of any form of existence, transitory or permanent, that is limited or determined by conditions.

What, then, is the origin of the word 'reality'? This comes from the Latin 'res', which means 'thing'. To be real is to be a 'thing'. 'Reality' in its earlier meaning would then signify 'thinghood in general' or 'the quality of being a thing'.

It is particularly interesting that 'res' comes from the verb 'reri', meaning 'to think', so that literally, 'res' is 'what is thought about'. It is of course implicit that what is thought about has an existence that is independent of the process of thought, or in other words, that while we create and sustain an idea as a mental image by thinking about it, we do not create and sustain a 'real thing' in this way. Nevertheless, the 'real thing' is limited by conditions that can be expressed in terms of thought. Of course, the real thing has more in it than can ever be implied by the content of our thought about it, as can always be revealed by further observations. Moreover, our thought is not in general completely correct, so that the real thing may be expected ultimately to show behaviour or properties contradicting some of the implications of our thought about it. These are, indeed, among the main ways in which the real thing can demonstrate its basic independence from thought. The main indication of the relationship between thing and thought is, then, that when one thinks correctly about a certain thing, this thought can, at least up to a point, guide one's actions in relationship to that thing to produce an overall situation that is harmonious and free of contradiction and confusion.

If the thing and the thought about it have their ground in the one undefinable and unknown totality of flux, then the attempt to explain their relationship by supposing that the thought is in reflective correspondence with the thing has no meaning, for both thought and thing are forms abstracted from the total process. The reason why these forms are related could only be in the ground from which they arise, but there can be no way of discussing reflective correspondence in this ground, because reflective correspondence implies knowledge, while the ground is beyond what can be assimilated in the content of knowledge.

Does this mean that there can be no further insight into the relationship of thing and thought? We suggest that such further insight is in fact possible but that it requires looking at the question in a different way. To show the orientation involved in this way, we may consider as an analogy the well-known dance of the bees, in which one bee is able to indicate the location of honey-bearing flowers to other bees. This dance is probably not to be understood as producing in the 'minds' of the bees a form of knowledge in reflective correspondence with the flowers. Rather, it is an activity which, when properly carried out, acts as a pointer or indicator, disposing the bees to an order of action that will generally lead them to the honey. This activity is not separate from the rest of what is involved in collecting the honey. It flows and merges into the next step in an unbroken process. So one may propose for consideration the notion that thought is a sort of 'dance of the mind' which functions indicatively, and which, when properly carried out, flows and merges into an harmonious and orderly sort of overall process in life as a whole.

In practical affairs, it is fairly clear what this harmony and order mean (e.g., the community will be successful in producing food, clothing, shelter, healthy conditions of life, etc.), but man also engages in thought going beyond the immediately practical. For example, since time immemorial he has sought to understand the origin of all things and their general order and nature, in religious thought, in philosophy, and in science. This may be called thought that has 'the totality of all that is' as its content (for example, the attempt to comprehend the nature of reality as a whole). What we are proposing here is that such comprehension of the totality is not a reflective correspondence between 'thought' and 'reality as a whole'. Rather, it is to be considered as an art from, like poetry, which may dispose us toward order and harmony in the overall 'dance of the mind' (and thus in the general functioning of the brain and nervous system). This point has been made earlier, in the Introduction.

What is required here, then, is not an *explanation* that would give us some knowledge of the relationship of thought and thing, or of thought and 'reality as a whole'. Rather, what is needed is an *act of understanding*; in which we see the totality as an actual process that, when carried out properly, tends to bring about a harmonious and orderly overall action, incorporating both thought and what is thought about in a single movement, in which analysis into separate parts (e.g., thought and thing) has no meaning.

4 THOUGHT AND NON-THOUGHT

While it is thus clear that *ultimately* thought and thing cannot properly be analysed as separately existent, it is also evident that in man's immediate experience some such analysis and separation has to be made, at least provisionally, or as a point of departure. Indeed, the distinction between what is real and what is mere thought and therefore imaginary or illusory is absolutely necessary, not only for success in practical affairs but also if we are in the long run even to maintain our sanity.

It is useful here to consider how such a distinction may have arisen. It is well known,[3] for example, that a young child often finds it difficult to distinguish the contents of his thought from real things (e.g., he may imagine that these contents are visible to others, as they are visible to him, and he may be afraid of what others call 'imaginary dangers'). So while he tends to begin the process of thinking naïvely (i.e. without being explicitly conscious that he *is* thinking), at some stage he becomes consciously aware of the process of thought, when he realizes that some of the 'things' that he seems to perceive are actually 'only thoughts' and therefore 'no things' (or nothing) while others are 'real' (or something).

Primitive man must often have been in a similar situation. As he began to build up the scope of his practical technical thought in his dealings with things, such thought images became more intense and more frequent. In order to establish a proper balance and harmony in the whole of his life he probably felt the need to develop his thought about totality in a similar way. In this latter kind of thought, the distinction between thought and thing is particularly liable to become confused. Thus, as men began to think of the forces of nature and of gods, and as artists made realistic images of animals and gods, sensed as possessing magical or transcendent powers, man was led to engage in a kind of thought without any clear physical referent that was so intense, so unremittant, and so 'realistic' that he could no longer maintain a clear distinction between mental image and reality. Such experiences must eventually have given rise to a deeply felt urge to clear up this distinction (expressed in questions such as 'Who am I?', 'What is my nature?', 'What is the true relationship of man, nature and the gods?', etc.), for to remain permanently confused about what is real and what is not, is a state that man must ultimately find to be intolerable, since it not only makes impossible a rational approach to practical problems but it also robs life of all meaning.

It is clear, then, that sooner or later, man in his overall process of thought would engage in systematic attempts to clear up this distinction. One can see that at some stage it has to be felt in this process that it is not enough to know how to distinguish particular thoughts from particular things. Rather, it is necessary to understand the distinction universally. Perhaps, then, the primitive man or the young child may have a flash of insight in which he sees, probably without explicitly verbalizing it, that *thought as a whole* has to be distinguished from *the whole of what is not thought.* This may be put more succinctly as the distinction between thought and non-thought, and abbreviated further to T and NT. The line of reasoning implicit in such a distinction is:

> T is not NT (thought and non-thought are different and mutually exclusive).
>
> All is either T or NT (thought and non-thought cover the whole of what can exist).

In a certain sense, true thinking begins with this distinction. Before it is made, thinking may take place but, as indicated earlier, there can be no full consciousness that thinking is what is taking place. So, thought proper begins in this way with thought, conscious of itself through its distinguishing itself from non-thought.

Moreover, this step in which thought proper begins is perhaps man's first thought with the totality as its content. And we can see how deeply such thought is embedded in the consciousness of all mankind, and how it arises very early as a necessary stage in the attempt of thought to bring sanity and order to its 'dance'.

This mode of thought is further developed and articulated by trying to discover various distinguishing characteristics or qualities that belong to thought and to non-thought. Thus, non-thought is commonly identified with reality, in the sense of thinghood. As indicated earlier, real things are recognized mainly by their independence of how we think of them. Further characteristic distinctions are that real things may be palpable, stable, resistant to attempts to change them, sources of independent activity throughout the whole of reality. On the other hand, thoughts may be regarded as mere 'mental stuff', impalpable, transient, easily changed, and not capable of initiating independent lines of activity outside of themselves, etc.

Ultimately, however, such a fixed distinction between thought and non-thought cannot be maintained, for one can see that thought is a real activity, which has to be grounded in a broader totality of real movement and action that overlaps and includes thought.

Thus, as has already been pointed out, thought is a material process whose content is the total response of memory, including feelings, muscular reactions and even physical sensations, that merge with and flow out of the whole response. Indeed, all man-made features of our general environment are, in this sense, extensions of the process of thought, for their shapes, forms, and general orders of movement originate basically in thought, and are incorporated within this environment, in the activity of human work, which is guided by such thought. Vice versa, everything in the general environment has, either naturally or through human activity, a shape, form, and mode of movement, the content of which 'flows in' through perception, giving rise to sense impressions which leave memory traces and thus contribute to the basis of further thought.

In this whole movement, content that was originally in memory continually passes into and becomes an integral feature of the environment, whole content that was originally in the environment passes into and becomes an integral feature of memory, so that (as pointed out earlier) the two participate in a single total process, in which analysis into separate parts (e.g. thought and thing) has ultimately no meaning. Such a process, in which thought (i.e. the response of memory) and the general environment are indissolubly linked, is evidently of the nature of a *cycle*, as illustrated symbolically in figure 3.1. (though of course the cycle should be regarded more accurately as always opening out into a spiral). This cyclical (or spiral) movement, in which thought

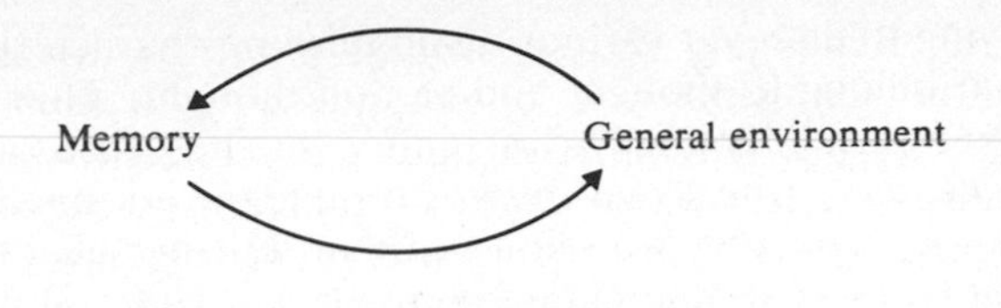

Figure 3.1

has its full actual and concrete existence, includes also the communication of thoughts between people (who are parts of each other's environment) and it goes indefinitely far into the past. Thus, at no stage can we properly say that the *overall process* of thought begins or ends. Rather, it has to be seen as one unbroken totality of movement, not belonging to any particular person, place, time, or group of people. Through the consideration of the physical nature of the response of memory in reactions of nerves, feelings, muscular motions, etc., and through the consideration of the merging of these responses with the general environment in the overall cyclical process described above, we see then that thought *is* non-thought (T *is* NT).

Vice versa, however, we can see also that non-thought *is* thought (NT *is* T). Thus, 'reality' *actually is* a word with a certain implied thought content. To be sure, this may be said of any term in our language, but, as has been seen, such terms may generally indicate real things, which we can in principle perceive. There is no way, however, to look at reality as if it were some sort of 'thing', in order to test whether our idea fits or does not fit this 'thing called reality'. We have indeed already suggested in this connection that the term 'reality' indicates an unknown and undefinable totality of flux, that is the ground of all things and of the process of thought itself, as well as of the movement of intelligent perception. But this does not basically alter the question, for if reality is thus unknown and unknowable, how can we be certain that it is there at all? The answer is, of course, that we can't be certain.

Nevertheless, it does not follow from this that 'reality' is a meaningless word, for, as we have already seen, the mind in its 'dance of thought' can in the long run move in an orderly and sane manner only if the 'form of the dance' includes some sort of distinction between thought and non-thought (i.e., reality). We have also seen, however, that this distinction has to be made in the ever-changing flux of process in which thought passes into non-thought while non-thought passes into thought, so that it cannot be regarded as fixed. Such a non-fixed distinction evidently requires the free movement of intelligent perception, which can, on each occasion, discern what content originates in thought and what content originates in a reality that is independent of thought.

It is clear, then, that the term 'reality' (which in this context means 'reality as a whole') is not properly to be regarded as part of the content of thought. Or, to put this in another way, we may say that *reality is no thing* and that is also not the *totality of all things* (i.e., we are not to identify 'reality' with 'everything'). Since the word 'thing' signifies a conditioned form of existence, this means that 'reality as a whole' is also not to be regarded as conditioned. (Indeed, it could not consistently be so regarded, because the very term 'reality as a whole' implies that it contains all factors that could condition it and on which it could depend.) So any notion of totality based on a fixed and permanent distinction between thought and reality must collapse when applied to the totality.

The original form of the fixed distinction between thought and reality (i.e., non-thought) was:

T is not NT
All is either T or NT

This form is characteristic of what is called Aristotelean logic (though of course it is probably as old as thought proper, while Aristotle was merely the earliest person known to us who enunciated it clearly and succinctly). This may be called the logic proper to things. Any particular thought form that fits this logic can, of course, be applicable to a corresponding thing only under certain conditions which are required for that thing to be what it is. That is to say, a set of thought forms that follow the rules of Aristotelean logic will service as adequate guides in activities incorporating things only in some limited domain beyond which these things must change or behave in new ways, so that other thought forms will then be needed.

When we come to consider the 'totality of all that is', however, our primary concern is, as we have seen, not with conditioned things but with the unconditioned totality that is the ultimate ground of all. Here, the rules enunciated by Aristotle break down, in the sense that there is not even a limited domain or set of conditions under which they could apply: for, in addition to the Aristotelean rules, we have to assert the following:

> T *is* NT
> NT *is* T
>
> All is *both* T and NT (i.e., the two merge and flow into each other, in a single unbroken process, in which they are ultimately one).
>
> All is *neither* T *nor* NT (i.e., the ultimate ground is unknown, and therefore not specifiable, neither as T nor NT, nor in any other way).

If the above is combined with the original 'T is not NT' and 'All is either T or NT', and if we further suppose that 'T' and 'NT' are names of *things*, we will imply absolute self-contradiction. What we are doing here is to regard this whole combination as an indication that 'T' and 'NT' are not names of things. Rather, as indicated earlier, they are to be considered as terms in our discourse whose function is to dispose the mind to an act of intelligent perception, in which what is called for is to discern, in each case, what content originates in thought (i.e., the response of memory) and what content originates in some 'reality' that is independent of thought. Since the reality that is independent of thought is ultimately unknown and unknowable, such a discernment evidently cannot take the form of assigning a particular feature of the content to a particular fixed category, T or NT. Rather, if there is an awareness of the ever-changing *totality*, of what originates in thought (i.e., in the response of memory, which is the field of the known), then, by implication, whatever is *not* in this totality has to be treated as originating independent of thought.

It is clearly extremely important that no part of what originates in the response of memory be missed or left out of awareness. That is to say, the primary 'mistake' that can be made in this field is not the *positive* one of wrongly assigning what originates in thought to a reality independent of thought. Rather, it is the *negative* one of overlooking or failing to be aware that a certain movement originates in thought, and thus implicitly treating that movement as originating in non-thought. In this way, what is actually the one single process of thought is tacitly treated as if it were split in two parts (but of course without one being aware that this is happening). Such unconscious fragmentation of the process of thought must lead to distortion in all of perception, one's own responses of memory to a reality that would be independent of these responses, there will be a further 'feedback' leading to more irrelevant thoughts about this 'independent reality', and these thoughts will constitute yet further inappropriate responses of memory which add to this 'independent reality' in a self-maintaining process that is generally very hard to break out of. This kind of feedback (which we have indicated earlier in connection with the analogy in which thought is regarded as similar to a radio receiver) will eventually tend to confuse the entire operation of the mind.

5 THE FIELD OF KNOWLEDGE, CONSIDERED AS PROCESS

In ordinary experience, in which we deal with sensuously perceivable things, it is usually possible sooner or later for intelligent perception to discern clearly the totality of those aspects of experience originating in thought (and by implication the totality of those originating independently of thought). However, as we have seen, in thought that aims to have totality as its content it is much harder to have such clarity, on the one hand because this thought is so intense, continuous and total that it gives a strong impression of reality, and on the other hand because there are no sensuously perceivable 'things' against which it could be tested. It is thus quite easy, through inadequate attention to the actual process of one's thought, to 'slip into' a form of conditioned response of memory, in which one is not alert to the fact that is still only a form of thought, a form that aims to give a view of 'the whole of reality'. So, 'by default' one falls into the trap of tacitly treating such a view as originating independently of thought, thus implying that its content *actually* is the whole of reality.

From this point on, one will see, in the whole field accessible to one, no room for change in the overall order, as given by one's notions of totality, which indeed must now seem to encompass all that is possible or even thinkable. This means, however, that our knowledge about 'the whole of reality' will then have to be regarded as having a fixed and final form, which reflects or reveals a correspondingly fixed and final form of what this total reality actually is. To adopt such an attitude will evidently tend to prevent that free movement of the mind needed for clarity of perception, and so will contribute to a pervasive distortion and confusion, extending into every aspect of experience.

As indicated earlier, thought with totality as its content has to be considered as an art form, like poetry, whose function is primarily to give rise to a new perception, and to action that is implicit in this perception, rather than to communicate reflective knowledge of 'how everything is'. This implies that there can no more be an ultimate form of such thought than there could be an ultimate poem (that would make all further poems unnecessary).

Any particular form of thinking about the totality does indeed indicate a way of looking at our whole contact with reality, and thus it has implications for how we may act in this contact. However, each such way of looking is limited, in the sense that it can lead to overall order and harmony only up to some point, beyond which it must cease to be relevant and fitting. (Compare with the notion of truth in function in chapter 2.) Ultimately, the actual movement of thought embodying any particular notion of totality has to be seen as a process, with ever-changing form and content. If this process is carried out properly, with attention to and awareness of thought in its actual flux of becoming, then one will not fall into the habit of treating the content tacitly as a final and essentially static reality that would be independent of thought.

Even this statement about the nature of our thinking is, however, itself only a form in the total process of becoming, a form which indicates a certain order of movement of the mind, and a certain disposition needed for the mind to engage harmoniously in such movement. So there is nothing final about it. Nor can we tell where it will lead. Evidently, we have to be open to further fundamental changes of order in our thought as we go on with the process. Such changes have to come about in fresh and creative acts of insight, which are necessary for the orderly movement of such thought. What we are suggesting in this chapter, is, then, that only a view of knowledge as an integral part of the total flux of process may lead generally to a more harmonious and orderly approach to life as a whole rather than a static and fragmentary view, which does not treat knowledge as process, and which splits knowledge off from the rest of reality.

It is important in this context to emphasize that permanently to identify certain views concerning the totality as belonging to Whitehead, or to someone else, is to interfere with treating knowledge *consistently* as an integral part of an overall process. Indeed, whoever takes up Whitehead's views is actually taking these as a point of departure, in a further process of the *becoming of knowledge*. (Perhaps we could say that he is working further down the 'stream of knowledge'.) In this process, some aspects may change fairly slowly, while others change more rapidly, but the key point to keep in mind is that the process has no definable aspect that is *absolutely* fixed. Intelligent perception is of course needed, for moment to moment, to discern those aspects that should properly change slowly and those that should properly change rapidly, as one works in the 'art form' of creation of ideas about 'the totality of all that is'.

We have to be very alert and careful here, for we tend to try to fix the essential content of our discussion in a particular concept or image, and talk about this as if it were a separate 'thing' that would be independent of our thought about it. We fail to notice that in fact this 'thing' has by now become only an image, a form in the overall process of thought, i.e., response of memory, which is a residue of past perception through the mind (either someone else's or one's own). Thus, in a very subtle way, we may once again be trapped in a movement in which we treat something originating in our own thought as if it were a reality originating independently of this thought.

We can keep out of this trap by being aware that the actuality of knowledge is a living process that is taking place *right now* (e.g. in this room). In such an actual process, we are not just talking about the movement of knowledge, as if looking at it from the outside. We are actually taking part in this movement, and are aware that this is indeed what it is happening. That is to say, it is a genuine reality for all of us, a reality which we can observe and to which we can give our attention.

The key question is, then: 'Can we be aware of the ever-changing and flowing reality of this actual process of knowledge?' If we can think from such awareness, we will not be led to mistake what originates in thought with what originates in reality that is independent of thought. And thus, the art of thinking with totality as its content may develop in a way that is free of the confusion inherent in those forms of thought which try to define, once and for all, what 'the whole of reality is', and which therefore lead us to mistake the content of such thought for the overall order of a total reality that would be independent of thought.

4

Hidden variables in the quantum theory

The question of whether there are hidden variables underlying the quantum theory was thought to have been settled definitely in the negative long ago. As a result, the majority of modern physicists no longer regard this question as relevant for physical theory. In the past few years, however, a number of physicists, including the author, have developed a new approach to this problem, which raises the question of hidden variables again.[1] It is my purpose here to review briefly the main features of what has been accomplished thus far in this new approach, and therefore to indicate some general lines on which theories involving hidden variables are currently developing.

In the course of this chapter, we shall show a number of reasons why theories involving hidden variables promise to be significant for the treatment of new physical problems, especially those arising in the domain of very short distances (of the order of 10^{-13}cm or less) and of very high energies (of the order of 10^9 ev or more). Finally, we shall answer the main objections that have been raised against the notion of hidden variables; i.e. the difficulties of dealing with the Heisenberg indeterminacy relations, the quantization of action, the paradox of Einstein, Rosen and Podolsky, and von Neumann's arguments against the possibility of such variables.

1 MAIN FEATURES OF THE QUANTUM THEORY

In order to understand the way the theory of hidden variables has developed, it is first of all necessary to keep clearly in mind the main features of the quantum theory. Although there are several alternative formulations of this theory (due to Heisenberg, Schrödinger, Dirac, von Neumann, and Bohr), which differ somewhat in interpretation,[2] they all have the following basic assumptions in common:

1 The fundamental laws of the quantum theory are to be expressed with the aid of a *wave function* (in general, many dimensional), which satisfies a linear equation (so that solutions can be superposed linearly).

2 All physical results are to be calculated with the aid of certain 'observables', represented by Hermitian operators, which operate linearly on the wave function.

3 Any particular observable is definite (sharply defined) only when the wave function is an eigenfunction of the corresponding operator.

4 When the wave function is not an eigenfunction of this operator, then the result of a measurement of the corresponding observable cannot be determined beforehand. The results of a series of measurements on an ensemble of systems represented by the same wave function will fluctuate at random (lawlessly) from one case to the next, over the various possibilities.

5 If the wave function is given by

$$\psi = \sum_n C_n \psi_n$$

where ψ_n is the eigenfunction of the operator in question corresponding to the nth eigenvalue, the probability of obtaining the nth eigenvalue in a large ensemble of measurements will be given by $P_n = |C_n|^2$.

6 Because of the non-commutation of many operators (such as p and x) which correspond to variables that must be defined together in classical mechanics, it follows that no wave functions can exist which are simultaneous eigenfunctions of all the operators that are significant for a given physical problem. This means that not all physically significant observables can be determined together, and even more important, that those which are not determined will fluctuate lawlessly (at random) in a series of measurements on an ensemble represented by the same wave function.

2 LIMITATIONS ON DETERMINISM IMPLIED BY THE QUANTUM THEORY

From the features described in the previous section, one sees immediately that there exists a certain limitation on the degree to which the results of *individual* measurements are determined according to the quantum theory. This limitation applies to any measurement that depends appreciably on the quantum properties of matter. Thus, in an ensemble of radioactive nuclei, the decay of each nucleus can be detected *individually* by the click of a Geiger counter. A more detailed study of the quantum mechanics of the problem shows that the operator corresponding to the measurement of a decay product does not commute with the operator whose eigenfunctions represent the undisintegrated. Thus it follows that if we begin with an ensemble of undisintegrated nuclei, represented by the same wave function, each individual nucleus will decay at an unpredictable time. This time will vary from one nucleus to another in a lawless way, while only the mean fraction that decays in a given interval of time can be predicted approximately from the wave function. When such predictions are compared with experiment, it is indeed discovered that there is a random distribution of clicks of the Geiger counter, together with a regular mean distribution that obeys the probability laws implied by the quantum theory.

3 ON THE INTERPRETATION OF INDETERMINISM IN THE QUANTUM THEORY

From the fact that quantum theory agrees with experiment in so wide a domain (including the problem treated in the previous section as a special but typical case), it is evident that the indeterministic features of quantum mechanics are in some way a reflection of the real behaviour of matter in the atomic and nuclear domains, but here the question arises as to just how to interpret this indeterminism.

To clarify the meaning of this question, we shall consider some analogous problems. Thus, it is well known that insurance companies operate on the basis of certain statistical laws, which predict to a high degree of approximation the mean number of people in a given class of age, height, weight, etc., that will die of a certain disease in a specified period of time. They can do this even though they cannot predict the precise time of death of an individual policy holder, and even though such individual deaths are distributed at random in a way having no lawful relationship to the kind of data that the insurance company is able to collect. Nevertheless, the fact that statistical laws of this kind are operating does not prevent the simultaneous operation of individual laws which determine in more detail the precise conditions of death of each policyholder (e.g., a man may cross a road at a particular time and be struck by a car, he may be exposed to disease germs while he is in a weak state, etc.), for when the same result (death) can be produced by a large number of essentially independent causes, there is no reason why these causes should not be distributed in just such a way as to lead to statistical laws in a large aggregate.

The importance of such considerations is quite evident. Thus, in the field of medical research, the operation of statistical laws is never regarded as a reason against the search for more detailed individual laws (e.g., as to what makes a given individual die at a given time, etc.).

Similarly, in the field of physics, when it was discovered that spores and smoke particles suffer a random movement obeying certain statistical laws (the Brownian motion) it was supposed that this was due to impacts from myriads of molecules, obeying deeper individual laws. The statistical laws were then seen to be consistent with the possibility of deeper individual laws, for, as in the case of insurance statistics, the overall behaviour of an individual Brownian particle would be determined by a very large number of essentially independent factors. Or, to put the case more generally: *lawlessness of individual behaviour in the context of a given statistical law is, in general, consistent with the notion of more detailed individual laws applying in a broader context.*

In view of the above discussion, it seems evident that, at least on the face of the question, we ought to be free to consider the hypothesis that results of individual quantum-mechanical measurements are determined by a multitude of new kinds of factors, outside the context of what can enter into the quantum theory. These factors would be represented mathematically by a further set of variables, describing the states of new kinds of entities existing in a deeper, sub-quantum-mechanical level and obeying qualitatively new types of individual laws. Such entities and their laws would then constitute a new side of nature, a side that is, for the present 'hidden'. But then the atoms, first postulated to explain Brownian motion and large-scale regularities, were also originally 'hidden' in a similar way, and were revealed in detail only later by new kinds of experiments (e.g., Geiger counters, cloud chambers, etc.) that were sensitive to the properties of individual atoms. Similarly, one may suppose that the variables describing the sub-quantum-mechanical entities will be revealed in detail when we have discovered still other kinds of experiments, which may be as different from those of the current type as the latter are from experiments that are able to reveal the laws of the large-scale level (e.g., measurements of temperature, pressure, etc.).

At this point it must be stated that as is well known – the majority of modern theoretical physicists[3] have come to reject any suggestion of the type described above. They do this mainly on the basis of the conclusion that the statistical laws of the quantum theory are incompatible with the possibility of deeper individual laws. In other words, while they would in general admit that some kinds of statistical laws are consistent with the assumption of further individual laws operating in a broader context, they believe that quantum mechanics could never satisfactorily be regarded as a law of this kind. The statistical features of the quantum theory are thus regarded as representing a kind of irreducible lawlessness of *individual* phenomena in the quantum domain. All individual laws (e.g. classical mechanics) are then regarded as limiting cases of the probability laws of the quantum theory, approximately valid for systems involving large numbers of molecules.

4 ARGUMENTS IN FAVOUR OF THE INTERPRETATION OF QUANTUM-MECHANICAL INDETERMINISM AS IRREDUCIBLE LAWLESSNESS

We shall now consider the main arguments on which is based the conclusion that quantum-mechanical indeterminism represents a kind of irreducible lawlessness.

4.1 Heisenberg's indeterminacy principle

We begin with a discussion of Heisenberg's indeterminacy principle. He showed that even if one supposes that the physically significant variables actually existed with sharply defined values (as is demanded by classical mechanics) then we could never measure all of them simultaneously, for the interaction between the observing apparatus and what is observed always involves an exchange of one or more indivisible and uncontrollably fluctuating quanta. For example, if one tries to measure the coordinate, x, and the associated momentum, p, of a particle, then the particle is disturbed in such a way that the maximum accuracy for the simultaneous determination of both is given by the well-known relation $\Delta p \Delta x \geqslant h$. As a result, even if there were deeper sub-quantum laws determining the precise behaviour of an individual electron, there would be no way for us to verify by any conceivable kind of measurement that these laws were really operating. It is therefore concluded that the notion of a sub-quantum level would be 'metaphysical', or empty of real experimental content. Heisenberg argued that it is desirable to formulate physical laws in terms of the minimum possible number of such notions, for they add nothing to the physical predictions of the theory, while they complicate the expression in an irrelevant way.

4.2 Von Neumann's arguments against hidden variables

The next of the main arguments against hidden variables, i.e., that of von Neumann, will now be presented in a simplified form.

From postulates (4), (5) and (6) of section 1, it follows that no wave function can describe a state in which *all* physically significant quantities are 'dispersionless' (i.e., sharply defined and free from statistical fluctuation). Thus, if a given variable (say p) is fairly well defined, the conjugate variable (x) must fluctuate over a broad range. Let us suppose that, when the system is in such a state, there are hidden variables on a deeper level which determine just how x is going to fluctuate in each instance. Of course, we would have no need to determine the values of these hidden variables, and in a statistical ensemble of measurements of x, we would still obtain the same fluctuations as are predicted by the quantum theory. Nevertheless, each case that was going to give a certain value of x would belong to a certain set of values of the hidden variables, and as a result the ensemble could be regarded as made up of a corresponding set of distinct and clearly defined sub-ensembles.

Von Neumann argued, however, that such a set of distinct and clearly defined sub-ensembles is not consistent with certain other essential characteristics of the quantum theory, i.e., those associated with the *interference* between parts of the wave function corresponding to different values of x. To demonstrate this interference, we could refrain from measuring x but instead we do a third kind of measurement, which determines an observable that is sensitive to the form of the wave function over a wide region of space. For example, we could pass the particles through a grating and measure the diffraction pattern. (Von Neumann[4] actually discussed the case of an observable that corresponds to a sum of two or more non-commuting operators; but it is evident that in an interference experiment we realize physically an example of just such an observable, since the final result determines some complex combinations of position and momentum operators for the observed system.)

It is well known that in such an experiment a statistical interference pattern is still obtained, even if we pass the particles through the apparatus at intervals so far apart that each particle essentially enters separately and independently of all the others. But, if the whole ensemble of such particles were to split into sub-ensembles, each corresponding to the electron striking the grating at a definite value of x, then the statistical behaviour of every sub-ensemble would be represented by a state corresponding to a delta function of the point in question. As a result, a single sub-ensemble could have no interference that would represent the contributions from different parts of the grating. Because the electrons enter separately and independently no interference between sub-ensembles corresponding to different positions will be possible either. In this way we show that the notion of hidden variables is not compatible with the interference properties of matter, which are both experimentally observed and necessary consequences of the quantum theory.

Von Neumann generalized the above argument and made it more precise; but he came to essentially the same result. In other words, he concluded that nothing (not even hypothetical hidden variables) can be consistently supposed to determine beforehand the results of an individual measurement in more detail than is possible according to the quantum theory.

4.3 The paradox of Einstein, Rosen and Podolsky

The third important argument against hidden variables is closely connected with the analysis of the paradox of Einstein *et al.*[5] This paradox arose out of the point of view, originally rather widespread, of regarding the indeterminacy principle as *nothing more than* an expression of the fact that there is a minimum unpredictable and uncontrollable disturbance in every measurement process. Einstein, Rosen and Podolsky then suggested a hypothetical experiment, from which one could see the untenability of the above interpretation of Heisenberg's principle.

We shall give here a simplified form of this experiment.[6] Consider a molecule of zero total spin, consisting of two atoms of spin, $\hbar/2$. Let this molecule be disintegrated by a method not influencing the spin of either atom. The total spin then remains zero, even while the atoms are flying apart and have ceased to interact appreciably.

Now, if any component of the spin of one of the atoms (say A) is measured, then because the total spin is zero, we can immediately conclude that this component of the spin of the other atom (B) is precisely opposite. Thus, by measuring any component of the spin of the atom A, we can obtain this component of the spin of atom B, *without interacting with atom B in any way*.

If this were a classical system, no difficulties of interpretation would occur, because each component of the spin of each atom is always well defined and always remains opposite in value to the same component of the spin of the opposite atom. Thus the two spins are correlated and this permits us to know the spin of atom B when we measure that of A.

However, in the quantum theory we have the additional fact that only one component of the spin can be sharply defined at one time, while the other two are then subject to random fluctuations. If we wish to interpret the fluctuations as nothing but the result of disturbances due to the measuring apparatus, we can do this for atom A, which is directly observed, but how does atom B, which interacts in no way either with atom A or with the observing apparatus, 'know' in what direction it ought to allow its spin to fluctuate at random? The problem is made even more difficult if we consider that, while the atoms are still in flight, we are free to re-orientate the observing apparatus arbitrarily, and in this way to measure the spin of atom A in some other direction. This change is somehow transmitted *immediately* to atom B, which responds accordingly. Thus, we are led to contradict one of the basic principles of the theory of relativity, which states that no physical influences can be propagated faster than light.

The behaviour described above not only shows the untenability of the notion that the indeterminacy principle represents in essence only the effects of a disturbance due to the measuring apparatus; it also presents us with certain real difficulties if we wish to understand the quantum-mechanical behaviour of matter in terms of the notion of a deeper level of individual law operating in the context of a set of hidden variables.

Of course, if there are such hidden variables then they might perhaps be responsible for a 'hidden' interaction between atom B and atom A, or between atom B and the apparatus that measures the spin of atom A. Such an interaction, which would be over and above those that are explicitly taken into account in the quantum theory, could then, in principle, explain how atom B 'knows' what property of atom A is being measured; but the difficulty still remains that to explain the correlation for the case in which the apparatus was re-orientated while the atoms are still in flight, we would have to assume that this interaction is carried through space at a speed greater than that of light. This aspect of the problem is evidently one that any acceptable theory of hidden variables must somehow manage to deal with in a satisfactory way.

5 BOHR'S RESOLUTION OF THE PARADOX OF EINSTEIN, ROSEN AND PODOLSKY – THE INDIVISIBILITY OF ALL MATERIAL PROCESSES

The paradox of Einstein, Rosen and Podolsky was resolved by Niels Bohr in a way that retained the notion of indeterminism in quantum theory as a kind of irreducible lawlessness in nature.[7] To do this he used the *indivisibility* of a quantum as his basis. He argued that in the quantum domain the procedure by which we analyse classical systems into interacting parts breaks down, for whenever two entities combine to form a single system (even if only for a limited period of time) the process by which they do this is not divisible. We are therefore faced with a breakdown of our customary ideas about the indefinite analysability of each process into various parts, located in definite regions of space and time. Only in the classical limit, where many quanta are involved, can the effects of this indivisibility be neglected; and only there can we correctly apply the customary concepts of *detailed* analysability of a physical process.

To deal with this new property of matter in the quantum domain, Bohr proposed to begin with the classical level, which is immediately accessible to observation. The various events which take place in this level can be adequately described with the aid of our customary general concepts, involving indefinite analysability. It is then found that up to a certain degree of approximation these events are related by a definite set of laws, i.e., Newton's laws of motion, which would, in principle, determine the future course of these events in terms of their characteristics at a given time.

Now comes the essential point. In order to give the classical laws a real experimental content, we must be able to determine the momenta and positions of all the relevant parts of the system of interest. Such a determination requires that the system of interest be connected to an apparatus which yields some observable large-scale result that is definitely correlated to the state of the system of interest. But in order to satisfy the requirement that we must be able to know the state of the observed system by observing that of the large-scale apparatus, it must be possible, in principle at least, for us to distinguish between the two systems by means of a suitable conceptual analysis, even though they are connected and in some kind of interaction. In the quantum domain, however, such an analysis can no longer be correctly carried out. Consequently, one must regard what has previously been called the 'combined system' as a single, indivisible, overall *experimental situation*. The result of the operation of the whole experimental set-up does not tell us about the system that we wish to observe but, rather, only about itself as a whole.

The above discussion of the meaning of a measurement then leads directly to an interpretation of the indeterminacy relationships of Heisenberg. As a simple analysis shows, the impossibility of theoretically defining two non-commuting observables by a single wave function is matched exactly, and in full detail, by the impossibility of the operation together of two overall set-ups that would permit the simultaneous experimental determination of these two variables. This suggests that the non-commutativity of two operators is to be interpreted as a mathematical representation of the incompatibility of the arrangements of apparatus needed to define the corresponding quantities experimentally.

In the classical domain it is of course essential that pairs of canonically conjugate variables of the kind described above shall be defined together. Each one of such a pair describes a necessary aspect of the whole system, an aspect which must be combined with the other if the physical state of the system is to be defined uniquely and unambiguously. Nevertheless, in the quantum domain each one of such a pair, as we have seen, can be defined more precisely only in an experimental situation in which the other must become correspondingly less precisely defined. In a certain sense, each of the variables then opposes the other. Nevertheless, they still remain 'complementary' because each describes an essential aspect of the system that the other misses. Both variables must therefore still be used together but now they can be defined only within the limits set by Heisenberg's principle. As a result, such variables can no longer provide us with a definite, unique, and unambiguous concept of matter in the quantum domain. Only in the classical domain is such a concept in adequate approximation.

If there is no definite concept of matter in the quantum domain, what then is the meaning of the quantum theory? In Bohr's point of view it is just a 'generalization' of classical mechanics. Instead of relating to observable classical phenomena by Newton's equations, which are a completely deterministic and indefinitely analysable set of laws, we relate these same phenomena by the quantum theory, which provides a probabilistic set of laws that does not permit analysis of the phenomena in indefinite detail. The same concepts (e.g., position and momentum) appear in both classical and quantum theories. In both theories, all concepts obtain their experimental content in essentially the same way, i.e., by their being related to a specific experimental set-up involving observable large-scale phenomena. The only difference between classical and quantum theories is that they involve the use of different kinds of laws to relate the concepts.

It is evident that according to Bohr's interpretation nothing is measured in the quantum domain. Indeed, in his point of view, there can be nothing to measure there, because all 'unambiguous' concepts that could be used to describe, define, and think about the meaning of the results of such a measurement belong to the classical domain only. Hence, there can be no talk about the 'disturbance' due to a measurement, since there is no meaning to the supposition that there was something there to be disturbed in the first place.

It is now clear that the paradox of Einstein, Rosen and Podolsky will not arise, because the notion of some kind of actually existing molecule, which was originally combined, and which later 'disintegrated', and which was 'disturbed' by the 'spin-measuring' device, has no meaning either. Such ideas should be regarded as nothing more than picturesque terms that it is convenient to use in describing the whole experimental set-up by which we observe certain correlated pairs of classical events (e.g., two parallel 'spin-measuring' devices that are on opposite sides of the 'molecule' will always register opposite results).

As long as we restrict ourselves to computing the probabilities of pairs of events in this way, we will not obtain any paradoxes similar to that described above. In such a computation the wave function should be regarded as just a mathematical symbol, which will help us to calculate the right relationships between classical events, provided that it is manipulated in accordance with a certain technique, but which has no other significance whatsoever.

It is now clear that Bohr's point of view necessarily leads us to interpret the indeterministic features of the quantum theory as representing irreducible lawlessness; for, because of the indivisibility of the experimental arrangement as a whole, there is no room in the conceptual scheme for an ascription of causal factors which is more precise and detailed than that permitted by the Heisenberg relations. This characteristic then reveals itself as an irreducible random fluctuation in the detailed properties of the individual large-scale phenomena, a fluctuation, however, that still satisfies the statistical laws of the quantum theory. Bohr's rejection of hidden variables is therefore based on a very radical revision of the notion of what a physical theory is supposed to mean, a revision that in turn follows from the fundamental role which he assigns to the indivisibility of the quantum.

6 PRELIMINARY INTERPRETATION OF QUANTUM THEORY IN TERMS OF HIDDEN VARIABLES

In this section, we shall sketch the general outlines of certain proposals toward a specific new interpretation of the quantum theory, involving hidden variables. It must be emphasized at the outset that these proposals are only preliminary in form. Their main purpose is twofold: first, to point out in relatively concrete terms the meaning of some of our answers to the arguments against hidden variables that were summed up in the previous sections, and second, to serve as a definite starting point for the further and more detailed development of the theory, which will be discussed in later sections of this chapter.

The first systematic suggestions for an interpretation of the quantum theory in terms of hidden variables were made by the author.[8] Based at first on an extension and completion of certain ideas originally proposed by de Broglie,[9] this new interpretation was then carried further in a later work jointly by the author and Vigier.[10] After some additional development, it finally took a form the main points of which will be summarized as follows:[11]

1. The wave function, ψ, is assumed to represent an objectively real field and not just a mathematical symbol.

2. We suppose that there is, beside the field, a particle represented mathematically by a set of coordinates, which are always well defined and which vary in a definite way.

3. We assume that the velocity of this particle is given by

$$\vec{v} = \frac{\nabla S}{m} \tag{1}$$

where m is the mass of the particle, and S is a phase function, obtained by writing the wave function as $\psi = Re^{iS/\hbar}$, with R and S real.

4. We suppose that the particle is acted on not only by the classical potential V ($\mathbf{x}$) but also by an additional 'quantum potential',

$$U = -\frac{\hbar^2}{2m}\frac{\nabla^2 R}{R}. \tag{2}$$

5. Finally, we assume that the field ψ is actually in a state of very rapid random and chaotic fluctuation, such that the values of ψ used in the quantum theory are a kind of average over a characteristic interval of time, τ. (This time interval must be long compared with the mean periods of the fluctuations described above but short compared with those of quantum-mechanical processes.) The fluctuations of the ψ-field can be regarded as coming from a deeper sub-quantum-mechanical level, in much the same way that the fluctuations in the Brownian motion of a microscopic liquid droplet come from a deeper atomic level. Then, just as Newton's laws determine the mean behaviour of such a droplet, so Schrödinger's equation will determine the mean behaviour of the ψ-field.

On the basis of the above postulates, it is now possible to prove an important theorem, for, if the ψ-field fluctuates, then Eq. (1) implies that corresponding fluctuations will be communicated to the particle motion by the fluctuating quantum potential (2). Thus, the particle will not follow a completely regular trajectory but will have a track resembling that displayed in the usual kind of Brownian-motion particle. In this track there will be a certain *average* velocity given by an average of Eq. (1) over the field fluctuations occurring during the characteristic interval, τ. Then, on the basis of certain very general and reasonable assumptions concerning the fluctuations, which are described in detail elsewhere,[12] one can show that in its random motions the particle will spend the mean fraction of its time in the volume element, dV, of

$$P = |\psi|^2\, dV. \tag{3}$$

Thus, the field ψ is interpreted mainly as determining the motion through (1) and the 'quantum potential' through (2). The fact that it also determines the usual expression for the probability density then follows as a consequence of certain stochastic assumptions on the fluctuations of ψ.

It has been demonstrated[13] that the above theory predicts physical results that are identical with those predicted by the usual interpretation of the quantum theory, but it does so with the aid of very different assumptions concerning the existence of a deeper level of individual law.

To illustrate the essential differences between the two points of view, consider an interference experiment in which electrons of definite momentum are incident on a grating. The associated wave function ψ is then diffracted by the grating in relatively definite directions, and one obtains a corresponding 'interference pattern' from a statistical ensemble of electrons which have passed through the system.

As we saw in previous sections, the usual point of view does not permit us to analyse this process in detail, even conceptually; nor does it permit us to regard the places at which individual electrons will arrive as determined beforehand by the hidden variables. It is our belief, however, that this process can be analysed with the aid of a new conceptual model. This model is based, as we have seen, on the supposition that there is a particle following a definite but randomly fluctuating track, the behaviour of which is strongly dependent on an objectively real and randomly fluctuating ψ-field, satisfying Schrödinger's equation in the mean. When the ψ-field passes through the grating, it diffracts in much the same way as other fields would (e.g., the electromagnetic). As a result, there will be an interference pattern in the later intensity of the ψ-field, an interference pattern that reflects the structure of the grating. But the behaviour of the ψ-field also reflects the hidden variables in the sub-quantum level, which determine the details of its fluctuations around the mean value, otained by solving Schrödinger's equation. Thus, the place where each particle will arrive is finally determined in principle by a combination of factors including the initial position of the particle, the initial form of its ψ-field, the systematic changes of the ψ-field due to the grating, and the random changes of this field originating in the sub-quantum level. In a statistical ensemble of cases having the same mean initial wave function, the fluctuations of the ψ-field will, as has been shown,[14] produce just the same interference pattern that is predicted in the usual interpretation of the quantum theory.

At this point, we must ask how we have been able to come to

a result opposite to that deduced by von Neumann (section 4.2). The answer is to be found in a certain unnecessarily restrictive assumption behind von Neumann's arguments. This assumption has that the particles arriving at the grating in a given position, x (determined beforehand by the hidden variable), must belong to a sub-ensemble having the same statistical properties as those of an ensemble of particles whose position, x, has actually been measured (and whose functions are therefore all a corresponding delta function of position). Now it is well known that if the position of each electron as it passes through the grating were to be measured, no interference would be obtained (because of the disturbance due to the measurement that causes the system to divide into the non-interfering ensembles represented by delta functions as discussed in section 4.2). Hence, von Neumann's procedure is equivalent to an implicit assumption, that *any* factors (such as hidden variables) which determine x beforehand must destroy interference in the same way as it is destroyed in a measurement of the coordinate x.

In our model, we go beyond the above implicit assumption by admitting at the outset that the electron has more properties than can be described in terms of the so-called 'observables' of the quantum theory. Thus, as we have seen, it has a position, a momentum, a wave field, ψ, and sub-quantum fluctuations, all of which combine to determine the detailed behaviour of each individual system with the passage of time. As a result, the theory has room to describe within it the difference between an experiment in which the electrons pass through the grating undisturbed by anything else, and one in which they are disturbed by a position-measuring apparatus. These two sets of experimental conditions would lead to very different ψ-fields, even if in both cases the particles were to strike the grating at the same position. The differences in the subsequent behaviour of the electron (i.e., interference in one case and not in the other) will therefore follow from the different ψ-fields which exist in the two cases.

To summarize, we need not restrict ourselves to von Neumann's assumptions that sub-ensembles are to be classified only in terms of the values of quantum-mechanical 'observables'. Rather, such a classification must also involve further inner properties, at present 'hidden', which can later influence the directly observable behaviour of the system (as in the example we have discussed).

Finally, it is possible to study in a similar way how other characteristic problems are treated in terms of our new interpretation of the quantum theory (e.g., the Heisenberg inderminacy relation, and the paradox of Einstein, Rosen and Podolsky). This has in fact been done in some detail.[15] We shall, however, defer a discussion of these questions until after we have developed some additional ideas, because this will enable us to treat these problems in a simpler and clearer way than was possible previously.

7 CRITICISMS OF OUR PRELIMINARY INTERPRETATION OF QUANTUM THEORY IN TERMS OF HIDDEN VARIABLES

The interpretation of the quantum theory discussed in the previous section is subject to a number of serious criticisms.

First of all, it must be admitted that the notion of the 'quantum potential' is not entirely a satisfactory one, for not only is the proposed form, $U = -(\hbar^2/2m)(\nabla^2 R/R)$, rather strange and arbitrary but also (unlike other fields such as the electromagnetic) it has no visible source. This criticism by no means invalidates the theory as a logical self-consistent structure but only attacks its plausibility. Nevertheless, we evidently cannot be satisfied with accepting such a potential in a definitive theory. Rather, we should regard it as at best a schematic representation of some more plausible physical idea to which we hope to advance later, as we develop the theory further.

Second, in the many-body problem, we are led to introduce a many-dimensional ψ-field $[\psi(x_1, x_2, \ldots, x_n, \ldots, xN)]$ and a corresponding many-dimensional quantum potential

$$U = -\frac{\hbar^2}{2m} \sum_{i=1}^{N} \frac{\nabla_i^2 R}{R},$$

with $\psi = Re^{is/\hbar}$ as in the one-body case. The momentum of each particle is then given by

$$P_i = \frac{\partial S(x_1 \ldots x_n \ldots x_N)}{\partial x_i}. \tag{4}$$

All of these notions are quite consistent logically. Yet it must be admitted that they are difficult to understand from a physical point of view. As best, they should be regarded, like the quantum potential itself, as schematic or preliminary representations of certain features of some more plausible physical ideas to be obtained later.

Third, the criticism has been levelled against this interpretation that the precise values of the fluctuating ψ-field and of the particle coordinates are empty of real physical content. The theory has been constructed in just such a way that the observable large-scale results of any possible kind of measurements are identical with those predicted by current quantum theory. In other words, from the experimental results, one can find no evidence for the existence of the hidden variables, nor does the theory permit their definition to be ever good enough to predict any result more accurately than the current quantum theory does.

The answer to this criticism must be considered in two contexts. First of all, it should be kept in mind that before this proposal was made there had existed a widespread impression that no conceptions of hidden variables at all, not even if they were abstract, and hypothetical, could possibly be consistent with the quantum theory. Indeed, to prove the impossibility of such a conception was the basic aim of von Neumann's theorem. Thus, to a considerable extent, the question had already been raised in an abstract way in certain aspects of commonly held formulations of the usual interpretation of the quantum theory. To show that it was wrong to throw out hidden variables because they could not be imagined, it was therefore sufficient to propose any logically consistent theory that explained the quantum mechanics, through hidden variables, no matter how abstract and hypothetical it might be. Thus, the existence of even a single consistent theory of this kind showed that whatever arguments one might continue to use against hidden variables, one could no longer use the argument that they are inconceivable. Of course, the specific theory that was proposed was not satisfactory for general physical reasons, but if one such theory is possible, then other and better theories may also be possible, and the natural implication of this argument is 'Why not try to find them?'

Secondly, to answer in full the criticism that these ideas are purely hypothetical we note that the logical structure of the theory makes room for the possibility of its being changed in such a way that it ceases to be completely identical with the current quantum mechanics in its experimental content. As a result, the details of the hidden variables (e.g. the fluctuations of the ψ-field and of the particle positions) will be able to reveal themselves in new experimental results not predicted by the quantum theory as it is now formulated.

At this point, one might perhaps raise the question as to whether such new results are even possible. After all, does not the general framework of the quantum theory already fit in with all known experimental results, and if so, how could there by any others?

To answer this question, we first point out that even if there existed no known experiments that the current quantum-theoretical framework failed to treat satisfactorily, the possibility would always still be open for new experimental results, not fitting into this framework. All experiments are necessarily done only in some limited domain, and even in this domain, only to a limited degree of approximation. Room is therefore always left open, logically speaking, for the possibility that when experiments are done in new domains and to new degrees of approximation, results will be obtained that do not fit completely into the framework of the current theories.

Physics has quite frequently developed in the way described above. Thus, Newtonian mechanics, thought originally to be of completely universal validity, was eventually found to be valid in a limited domain (velocity small compared with that of light) and only to a limited degree of approximation. Newtonian mechanics had to give way to the theory of relativity which utilized basic conceptions concerning space and time which were in many ways not consistent with those of Newtonian mechanics. The new theory was, therefore, in certain essential features qualitatively and fundamentally different from the old one. Nevertheless, within the domain of low velocities, the new theory approached the old one as a limiting case. In a similar way, classical mechanics eventually gave way to the quantum theory, which is very different in its basic structure, but which still contains classical theory as a limiting case, valid approximately in the domain of large quantum numbers. Agreement with experiments in a limited domain and to a limited degree of approximation is evidently no proof, therefore, that the basic concepts of a given theory have a completely universal validity.

From the above discussion we see that the experimental evidence taken by itself will always leave open the possibility of a theory of hidden variables that yields results differing from those of the quantum theory in new domains (and even in the old domains when carried to a sufficiently high degree of approximation). Now, however, we must have some more definite ideas as to which are the domains in which we expect the results to be new, and as to just what are the ways in which they ought to be new.

Here, we may hope to get some clues by considering problems in a domain where current theories do not yield generally satisfactory results, i.e. one connected with very high energies and very short distances. With regard to such problems, we first note that the present relativistic quantum field theory meets severe difficulties which raise serious doubts as to its internal self-consistency. There are the difficulties arising in connection with the divergences (infinite results) obtained in calculations of the effects of interactions of various kinds of particles and fields. It is true that for the special case of electromagnetic interactions such divergences can be avoided to a certain extent by means of the so-called 'renormalization' techniques. It is by no means clear, however, that these techniques can be placed on a secure logical mathematical basis.[16] Moreover, for the problem of mesonic and other interactions, the renormalization method does not work well even when considered as a purely technical manipulation of mathematical symbols, apart from the question of its logical justification. While it has not been proved conclusively, as yet, that the infinities described above are essential characteristics of the theory, there is already a considerable amount of evidence in favour of such a conclusion.[17]

It is generally agreed that, if as seems rather likely, the theory does not converge, then some fundamental change must be made in its treatment of interactions involving very short distances, from which domain all the difficulties arise (as one sees in a detailed mathematical analysis).

Most of the proponents of the usual interpretation of the quantum theory would not deny that such a fundamental change seems to be needed in the present theory. Indeed, some of them, including Heisenberg, are even ready to go so far as to give up completely our notions of a definable space and time, in connection with such very short distances, while comparably fundamental changes in other principles, such as those of relativity, have also been considered by a number of physicists (in connection with the theory of non-local fields). But there seems to exist a widespread impression that the principles of quantum mechanics almost certainly will not have to be changed in essence. In other words, it is felt that however radical the changes in physical theories may be they will only build upon the principles of the present quantum theory as a foundation, and perhaps enrich and generalize these principles by supplying them with a newer and broader scope of application.

I have never been able to discover any well-founded reasons as to why there exists so high a degree of confidence in the general principles of the current form of the quantum theory. Several physicists[18] have suggested that the trend of the century is away from determinism, and that a step backwards is not very likely. This, however, is a speculation of a kind that could easily be made in any period concerning theories that have hitherto been successful. (For example, classical physicists of the nineteenth century could have argued with equal justification that the trend of the times was toward *more* determinism, whereas future events would have proved this speculation wrong. Still others have adduced a psychological preference for indeterministic theories, but this may well be just a result of their having become accustomed to such theories. Classical physicists of the nineteenth century would surely have expressed an equally powerful psychological bias toward determinism.)

Finally, there is a widespread belief that it will not really be possible to carry out our suggested programme of developing a theory of hidden variables, which will be genuinely different in experimental content from the quantum theory, and which will still agree with the latter theory in the domain where this theory is already known to be essentially correct. This view is held in particular by Niels Bohr, who expressed especially strong doubts[19] that such a theory could treat all significant aspects of the problem of the *indivisibility* of the quantum of action – but then this argument stands or falls on the question of whether an alternative theory of the kind described above can really be produced, and in the next sections, we shall see that such a position is not a very secure one.

8 STEPS TOWARD A MORE DETAILED THEORY OF HIDDEN VARIABLES

From the discussion given in the previous section, it is clear that our central task is to develop a new theory of hidden variables. This theory should be quite different from the current quantum theory both in its basic concepts and in its general experimental content, and yet be capable of yielding essentially the same results as those of the current theory in the domain in which this latter theory has thus far been verified, and to the degree of approximation of the measurements that have actually been carried out. The possibility of distinguishing between the two theories experimentally will then arise either in new domains (e.g., very short distances) or in more accurate measurements carried out in the older domains.

Our basic starting point will be to try to provide a more concrete physical theory leading to ideas resembling those discussed in connection with our preliminary interpretation (section 6). In doing this, we must first recall that we have been regarding indeterminism as a real and objective property of matter, but one associated with a given limited context (in this case that of the variables of the quantum-mechanical level). We are supposing that in a deeper sub-quantum level, there are further variables which determine in more detail the fluctuations of the results of individual quantum-mechanical measurements.

Does the existing physical theory provide us with any hints as to the nature of these deeper sub-quantum-mechanical variables? To guide us in our search, we can begin by considering the current quantum theory in its most developed form, namely that of relativistic field theory. According to the principles of the current theory, it is essential that every field operator, ϕ_μ, be a function of a sharply defined point, $\mathbf{x}$, and that all interactions shall be between fields at the same point. This leads us to formulate our theories in terms of a non-denumerable infinity of field variables.

Of course, such a formulation must be made, even classically, but in classical physics one can assume that the fields vary *continuously*. As a result, one can effectively reduce the number of variables to a denumerable set (e.g., the average values of the fields in very small regions), essentially because the field changes over very short distances are negligibly small. As a simple calculation shows, however, this is not possible in the quantum theory, because the shorter the distances one considers, the more violent are the quantum fluctuations associated with the 'zero-point energy' of the vacuum. Indeed, these fluctuations are so large that the assumption that the field operators are continuous functions of positions (and time) is not valid in a strict sense.

Even in the usual quantum theory, the problem of a non-denumerable infinity of field variables presents several as yet unsolved basic mathematical difficulties. Thus, it is customary to deal with field theoretical calculations by starting with certain assumptions concerning the 'vacuum' state, and thereafter applying perturbation theory. Yet, in principle, it is possible to start with an infinite variety of very different assumptions for the vacuum state, involving the assignment of definite values to a set of completely discontinuous functions of the field variables, functions which 'fill' the space densely and yet leave a dense set of 'holes'. *These new states cannot be reached from the original 'vacuum' state by any canonical transformation*[20]. Hence they lead to theories that are, in general, different in physical content from those obtained with the original starting point. It is entirely possible that because of the divergences in field theoretical results, even the current renormalization techniques imply such an 'infinitely different' vacuum state; but even more important, it is necessary to stress that a reorganization of a non-denumerable infinity of variables usually leads to a different theory, and that the principles of such a reorganization will then be equivalent to basic assumptions about the corresponding new laws of nature.

Thus far, we have restricted the above discussion to the effects of a reorganization of a non-denumerable infinity of variables within the framework of the present quantum theory, but similar conclusions will hold even for a classical theory involving a non-denumerable infinity of variables. Thus, if we once give up the assumption of the *continuity* of the classical field, we see that there is the same scope for obtaining a different classical theory in such a reorganization as there is in the quantum theory.

At this point we ask ourselves whether it would ever be possible to reorganize a classical field theory in such a way that it becomes equivalent (at least in some approximation and within some domain) to the modern quantum field theory. In order to answer this question, we must evidently reproduce from the basic 'deterministic' law of our assumed non-denumerable infinity of 'classical' field variables, the fluctuations of quantum processes, the indivisibility of the quantum, and other essential quantum-mechanical properties, such as interference and the correlations associated with the paradox of Einstein, Rosen and Podolsky. It is with these problems that we shall concern ourselves in subsequent sections.

9 TREATMENT OF QUANTUM FLUCTUATIONS

Let us begin by assuming some 'deterministic' field theory. Its precise characteristics are unimportant for our purposes here. All that is important is to suppose the following properties.

1. There is a set of field equations which completely determines the changes of the field with time.

2. These equations are sufficiently non-linear to guarantee a

significant coupling between all wave components, so that (except perhaps in some approximation) solutions cannot be linearly superposed.

3. Even in the 'vacuum' the field is so highly excited that the mean field in each region, however small, fluctuates significantly, with a kind of turbulent motion that leads to a high degree of randomness in the fluctuations. *This excitation guarantees the discontinuity of the fields in the smallest regions.*

4. What we usually call 'particles' are relatively stable and conserved excitations on top of this vacuum. Such particles will be registered at the large-scale level, where all apparatus is sensitive only to those features of the field that will last a long time, but not to those features that fluctuate rapidly. Thus, the 'vacuum' will produce no visible effects at the large-scale level, since its fields will cancel themselves out on the average, and space will be effectively 'empty' for every large-scale process (e.g. as a perfect crystal lattice is effectively 'empty' for an electron in the lowest band, even though the space is full of atoms).

It is evident that there would be no way to solve such a set of field equations directly. The only possibility would be to try to deal with some kind of *average* field quantities (taken over small regions of space and time). In general, we could hope that a group of such average quantities would, at least within some approximation, *determine themselves* independently of the infintely complex fluctuations inside the associated regions of space.[21] To the extent that this happened, we could obtain approximate field laws, associated with a certain level of size, but these laws cannot be exact because the non-linearity of the equations means fields will necessarily be coupled in some way to the inner fluctuations that have been neglected. As a result, the mean fields will also fluctuate at random about their average behaviour. There will be a typical domain of fluctuation of the mean fields, determined by the character of the deeper field motions that have been left out. As in the case of the Brownian motion of a particle, this fluctuation will determine a probability distribution

$$dP = P(\phi_1, \phi_2, \ldots, \phi_k \ldots)\, d\phi_1\, d\phi_2 \ldots d\phi_k \ldots \qquad (5)$$

which yields the mean fraction of the time in which the variables, ϕ_1, $\phi_2 \ldots, \phi_k \ldots$, representing the mean fields in the regions, 1, 2 . . ., k . . . respectively, will be in the range $d\phi_1\, d\phi_2 \ldots d\phi_k \ldots$ (Note that P is in general a multidimensional function, which can describe statistical *correlations* in the field distributions.)

To sum up, we are reorganizing the non-denumerable infinity of field variables, and we are treating explicitly only some denumerable sets of these reorganized coordinates. We do this by defining a series of levels by average fields, each associated with a certain dimension, over which the averages are taken. Such a treatment can be justified only in those cases in which the denumerable sets of variables form a totality that, *within certain limits*, determines its own motions independently of the precise details of the non-denumerable infinity of coordinates that has necessarily been left out of account. Such self-determination is never complete, however, and its basic limits are defined by a certain minimum degree of fluctuation over a domain that depends on the coupling of the field coordinates in question to those that have been neglected. Thus we obtain a real and objective limitation on the degree of *self-determination* of a certain level, along with a probability function that represents the character of the statistical fluctuations which are responsible for the above described limitations on self-determination.

10 HEISENBERG'S INDETERMINACY PRINCIPLE

We are now ready to show how Heisenberg's indeterminacy principle fits into our general scheme. We shall do this by discussing the degree of determinism associated with a space-averaged field coordinate, ϕ_k, and the corresponding average of the canonically conjugate field momentum, π_k.

To simplify the discussion let us suppose that the canonical momentum is proportional to the time derivative of the field coordinate, $\partial\phi_k/\partial t$ (as is the case for many fields such as the electromagnetic, mesonic, etc.). Each such field coordinate fluctuates at random. This means that its instantaneous time derivative is infinite (as also happens in the case of the Brownian movement of a particle). As a result, there is no rigorous way to define such an instantaneous time derivative. Rather, we must discuss the average change of field, $\Delta\phi_k$, over a small region of time, Δt (just as we had to take the average also over a region of space). The average value of the field momentum over this time interval is then

$$\overline{\pi_k} = a\left(\frac{\Delta\phi_k}{\Delta t}\right) \tag{6}$$

where a is the constant of proportionality.

If the field fluctuates in a random way, then by the very definition of randomness, the region over which it fluctuates during the time, Δt is given by

$$\overline{(\delta\phi_k)^2} = b\,\Delta t \quad \text{or} \quad |\delta\phi_k| = b^{1/2}(\Delta t)^{1/2} \tag{7}$$

where b is another constant of proportionality, associated with the mean magnitude of the random fluctuations of the field.

Of course, the precise manner of fluctuation of the field is determined by the infinity of deeper field variables not taken into account, but in the context of the level in question, nothing determines this precise behaviour. In other words, $|\delta\phi_k|$ represents the maximum possible degree of determination of ϕ_k within the level of field quantities averaged over similar intervals of time.

From the definition (6), we see that π_k will also fluctuate at random over the range

$$\delta\pi_k = \frac{a\,|\,\delta\phi_k\,|}{\Delta t} = \frac{ab^{1/2}}{(\Delta t)^{1/2}} \;. \tag{8}$$

Multiplying (8) and (7) together, we obtain

$$\delta\pi_k\;\delta\phi_k = ab. \tag{9}$$

Thus the product of the maximum degree of determination of π_k *and that of* ϕ_k is a constant, ab, independent of the time interval, Δt.

It is immediately clear that the above result shows a strong analogy to Heisenberg's principle,[22] $\delta p \delta q \lesssim h$. The constant, ab, appearing in Eq. (9) plays the role that Planck's constant, h, plays in Heisenberg's principle. The universality of h therefore implies the universality of ab.

Now a is just a constant relating the field momentum to its time derivative and will evidently be a universal constant. The constant, b, represents the basic intensity of the random fluctuation. To suppose that b is a universal constant is the same as to assume that the random field fluctuations are at all places, at all times, and in all levels of size, essentially the same in character.

With regard to different places and times the assumption of the universality of the constant, b, is not at all implausible. The random field fluctuations (which here play a role similar to that of the 'zero-point' vacuum fluctuations in the usual quantum theory) are infintely large, so that any disturbances that might be made by further localized excitations or concentrations of energy occurring naturally, or produced in a laboratory experiment, would have a negligible influence on the general magnitudes of the basic random fluctuations. (Thus, the presence of matter as we know it on a large scale would mean the concentration of a non-fluctuating part of the energy, associated with a few extra grams per cubic centimetre on top of the infinite zero-point fluctuations of the 'vacuum' field.)

With regard to the problem of different levels of space and time intervals, however, the assumption of the universality of b is not so plausible. Thus, it is quite possible that the quantity b will remain constant for fields averaged over shorter and shorter time intervals only down to some characteristic time interval Δt_0, beyond which the quantity b may change. This is equivalent to the possibility that the degree of self-determination may not be limited by Planck's constant, h, for very short times (and for correspondingly short distances).

It is easy to suggest a theory having the characteristics described above. Thus, suppose that the 'zero-point' field fluctuations were in some kind of statistical equilibrium corresponding to an extremely high temperature, T. The mean fluctuation in the energy per degree of freedom would, according to the equipartition theorem, be of the order of κT, but this mean energy is also proportional to the mean of $(\partial\phi/\partial t)^2$ (as happens for example in a collection of harmonic oscillators). Thus, we write

$$\alpha \overline{\left(\frac{\partial \phi}{\partial t}\right)^2} = \kappa T = \frac{\alpha}{b^2} \overline{(\pi)^2} \tag{10}$$

where κ is Boltzmann's constant and α is a suitable constant of proportionality.

As a result, if the time interval, Δt, appearing in Eq. (8) is made shorter and shorter, it will not be possible for $(\pi)^2$ to increase without limit as is implied by Eq. (8) and (9). Rather, $(\pi)^2$ will stop increasing at some critical time interval defined by

$$\kappa T = \frac{\alpha}{b^2} \frac{a^2 b}{(\Delta t_0)^2}\,; \quad \text{or} \quad (\Delta t_0)^2 = \frac{\alpha a^2}{b \kappa T}\,. \tag{11}$$

For shorter time intervals (and correspondingly short distances) the degree of self-determination of the average fields would then not be limited precisely by Heisenberg's relations but instead by a weaker set of relations.

We have thus constructed a theory which contains Heisenberg's relations as a limiting case, valid approximately for fields averaged over a certain level of intervals of space and time. Nevertheless, fields averaged over smaller intervals are subject to a greater degree of self-determination than is consistent with this principle. From this, it follows that our new theory is able to reproduce, in essence at least, one of the essential features of the quantum theory, i.e. Heisenberg's principle and yet have a different content in new levels.

The question of how this new content of our theory could be revealed in experiments will be discussed in later sections. For the moment, we restrict ourselves to pointing out that the divergencies of present-day field theories are directly a result of contributions to the energy, charge, etc., coming from quantum fluctuations associated with infintely short distance and times. Our point of view permits us to assume that while the total fluctuation is still infinite, the fluctuation per degree of freedom ceases to increase without limit as shorter and shorter times are considered. In this way, field-theoretical calculations could be made to give finite results. Thus, it is clear already that divergences of the current quantum field theory may come from the extrapolation of the basic principles of this theory to excessively short intervals of time and space.

11 THE INDIVISIBILITY OF QUANTUM PROCESSES

Our next step is to show how quantization, i.e., the indivisibility of the quantum of action, fits into our notions concerning a sub-quantum-mechanical level. To do this, we begin by considering in more detail the problem of just how to define the field averages that are needed for the treatment of a non-denumerable infinity of variables. Here, we shall guide ourselves by certain results obtained in the very analogous many-body problem (e.g., the analysis of solids, liquids, plasmas, etc., in terms of their constituent atomic particles). In this problem, we are likewise confronted with the need to treat certain kinds of averages of deeper (atomic) variables. The totality of a set of such averages then determines itself in some approximation, while its details are subject to characteristic domains of random fluctuations arising from the lower-level (atomic) motions, in much the same way that was suggested for the averages of the non-denumerable infinity of field variables discussed in the previous sections.

Now, in the many-body problem, one deals with large-scale behaviour by working with *collective coordinates*,[23] which are an approximately self-determining set of symmetrical functions of the particle variables, representing certain overall aspects of the motions (e.g., oscillations). The collective motions are determined (within their characteristic domains of random fluctuation) by approximate *constants of the motion*. For the special but very widespread case that the collective coordinates describe nearly harmonic oscillations, the constants of the motion are the amplitudes of the oscillations and their initial phases. More generally, however, they may take the form of more complex functions of the collective coordinates.

It is often very instructive to solve for the collective coordinates by means of a canonical transformation. In classical mechanics,[24] this takes the form

$$P_k = \frac{\partial S}{\partial q_k} (q_1 \dots q_k \dots ; J_1 \dots J_n)$$

$$Q_n = \frac{\partial S}{\partial J_n} (q_1 \dots q_k \dots ; J_1 \dots J_n \dots)$$

where S is the transformation function, p_k and q_k are the momenta and one coordinates of the particles, and J_n and Q_n are the momenta of the collective degrees of freedom. Here, we suppose the J_n to be constants of the motion. In other words, we assume that the transformation is such that, at least in the domain in which the approximation of collective coordinates is a good one, the Hamiltonian is only a function of the J_n, and not of the Q_n. It then follows that the Q_n increase linearly with time so that they have the properties of the so-called 'angle-variables'.[25]

It is clear that a similar attack can be made on the problem of a non-denumerable infinity of field variables subject to a non-linear coupling with each other. To do this, we now let q_k, p_k represent the original canonically conjugate set of field variables, and we assume that there will be a set of overall large-scale motions which we represent by the constants of the motion, J_n and the canonically conjugate angle-variables, Q_n. It is clear that if such overall motions exist, they will manifest themselves relatively directly in high-level interactions, for by hypothesis, they are the motions that retain their characteristic features for a long time without being lost in the infinitely rapid random fluctuations, which average out to zero on a higher level.

Our next task is to show that the constants of the motion (which are, for harmonic oscillators, proportional to the energy of a large-scale collective degree of freedom) are quantized by the rule $J = nh$, where n is an integer, and h is Planck's constant. Such a demonstration will constitute an explanation of the wave-particle duality, since the collective degrees of freedom are already known to be wavelike motions, with harmonically oscillating amplitudes. In general, these waves will take the form of fairly localized packets, and if these packets have discrete and well-defined quantities of energy, momentum, and other properties, they will at the higher level, reproduce all the essential characteristics of particles. Yet they will have inner wavelike motions which will reveal themselves only under conditions in which there exist systems that can respond significantly to these finer details.

In order to show the quantization of the constants of the motion as described above, we first return to the preliminary interpretation of the quantum theory, given in sections 6 and 7. Here, we encountered a relation very similar to (12).

$$P_k = \frac{\partial S}{\partial q_k} \; (q_1 \ldots q_k \ldots).$$

The main difference between (4) and (12) is that the former does not contain any constants of the motion, whereas the latter does. But once the constants of the motion are specified, they are just numbers, which need only be given certain values which they thereafter retain. If this is done, the S of Eq. (12) will also no longer contain the J_n as explicitly represented variables. We can therefore regard the S of our preliminary interpretation, (4), as the actual S function, in which the constants of the motion have already been specified. S is then determined by the wave function, $\psi = Re^{is/\hbar}$. Thus, when we give the wave function, we define a transformation function $S = \hbar I_m (l_n \psi)$, which latter determines certain constants of the motion implicitly.

In order to see more clearly how the constants of the motion are determined by the S of Eq. (4) let us construct the *phase integral*

$$I_C = \sum_k \oint C p_k \delta q_k \,. \tag{14}$$

The integral is taken around some circuit C, representing a set of displacement, δq_k (virtual or real), in the configuration space of the system. If Eq. (13) applies, we then obtain

$$I_C = \oint \sum_k \frac{\partial S}{\partial q_k} \delta q_k = \delta S_C \tag{15}$$

where δS_c is the change of S in going around the circuit C.

It is well known that the I_c, which are the so-called 'action variables' of classical mechanics, generally represent the constants of the motion. (For example, in the case of a set of coupled oscillators, harmonic or otherwise, the basic constants of the motion can be obtained by evaluating the I_c with suitably defined circuits.)[26] The wave function, ψ, which defines a certain function, S, therefore implies a corresponding set of constants of the motion.

Now, according to the current quantum theory, the wave function, $\psi = Re^{iS/h}$, is a single-valued function of all its dynamical coordinates, q_k. Thus, we must have

$$\delta S_c = 2\eta\pi\hbar = nh \tag{16}$$

where n is an integer.

The actual functions, S, obtained from the wave function, ψ, therefore imply that the basic constants of the motion for the system are discrete and quantized.

If the integer, n, is not zero, then as a simple calculation shows, there must be a discontinuity somewhere inside the circuit. But since $S = \hbar I_m (l_n \psi)$, and since ψ is a continuous function, a discontinuity of S will generally occur where ψ (and therefore R^2) has a zero. As we shall see presently, R^2 is the probability density for the system to be at a certain point in configuration space. The system will therefore have no probability of being at a zero of ψ, with the result that the singularities of S will imply no inconsistencies in the theory.

In many ways, the quantization described above resembles the old Bohr-Sommerfeld rule; yet it is basically different in its meaning. Here, the action variable, I_c, that is quantized is not obtained by using the simple expression of classical mechanics for the p_k in Eq. (14). Rather, it is obtained by using the expression (12), which involves the transformation functions, S, a function that depends on the non-denumerable infinity of variables, q_k. In certain sense, we can say that the old Bohr-Sommerfeld rule would be exactly correct if it were made to refer to the non-denumerable infinity of field variables, instead of just to the values of the variables that one obtains by solving the simple classical equations of motion for a small number of abstracted coordinates, Q_n.

Before going ahead to suggest an explanation of why δS_c should be restricted to the discrete values denoted by Eq. (16), we shall sum up and develop in a systematic way the main physical ideas to which we have thus far been led.

1. We abstract from the non-denumerable infinity of variables a set of 'collective' constants of the motion, J_n and their canonically conjugate quantities, Q_n.

2. The J_n can be consistently restricted to discrete integral multiples of h. Thus, action can be quantized.

3. If this set of coordinates determined itself completely, the Q_n would (as happens in typical classical theories) increase linearly with time. However, because of fluctuations due to the variables left out of the theory, the Q_n will fluctuate at random over the range accessible to them.

4. This fluctuation will imply a certain probability distribution of the Q_n having a dimensionality equal to 1 per degree of freedom (and not 2, as is the case for typical *classical* statistical distributions in phase space). When this distribution is transformed to the configuration space of the q_k there will be a corresponding probability function, $p(q_1 \ldots q_k \ldots)$, which also has a dimensionality of 1 per degree of freedom (the momenta, p_k being always determined in terms of q_k by Eq. (12)).

5. We then interpret the wave function $\psi = Re^{is/\hbar}$, by setting $p(q_1 \ldots q_k \ldots) = R^2(q_1 \ldots q_k \ldots)$ and by letting S be the transformation function that defines the constants of the motion of the system. It is clear that we have in this way given the wave function a meaning quite different from the one that was suggested in the preliminary interpretation of section 5, even though the two interpretations stand in a fairly definite relation to each other.

6. Because of the effects of the neglected lower-level field variables, the quantities I_n will, in general, remain constant only for some limited period of time. Indeed, as the wave function changes, the integral around a given circuit, $\Sigma_k \oint_c p_k \delta q_k = \delta S_c$ will change abruptly, whenever a singularity of S (and therefore

a zero of ψ) crosses the circuit, C. Hence discrete changes, by some multiple of h will occur in the action variables for non-stationary states.

12 EXPLANATION OF QUANTIZATION OF ACTION

In the previous section, we developed a theory involving a non-denumerable infinity of field variables that *has room* for the quantization of action according to the usual rules of the quantum theory. We shall now suggest a more definite theory, which will give possible physical reasons explaining why action is quantized by the rules described above, and which will show possible limitations on the domain of validity of these rules.

Our basic problem evidently is to propose some direct physical interpretation of the function, S which appears in the phase of the wave function (as $\psi = Re^{is/\hbar}$) and which is, according to our theory, also the transformation function defining the basic constants of the motion (see Eq. (15)); for if we are to explain why the change of S around a circuit is restricted to discrete multiples of h we must evidently assume that S is somehow related to some physical system, in such a way that $e^{is/\hbar}$ cannot be other than single-valued.

To give S a physical meaning that leads to the property described above, we shall begin with certain modifications of an idea originally suggested by de Broglie.[27] Let us suppose that the infinity of non-linearly coupled field variables is in reality so organized that in each region of space and time associated with any given level of size there is taking place a periodic inner process. The precise nature of this process is not important for our discussion here, as long as it is periodic (e.g., it could be an oscillation or a rotation). This periodic process would determine a kind of inner time for each region of space, and it would therefore effectively constitute a kind of local 'clock'.

Now every localized periodic process has, by definition, some Lorentz frame in which it remains at rest, at least for some time (i.e., in which it does not significantly change its mean position during this time). We shall further assume that, in this frame, neighbouring clocks of the same level of size will tend to be nearly at rest. Such an assumption is equivalent to the requirement that, in every level of size, the division of a given region into small regions, each containing its effective clock, has a certain regularity and permanence, at least for some time. If these clocks are considered in another frame (e.g., that of the laboratory), every effective clock will then have a certain velocity, which can be represented by a continuous function $\mathbf{v}(\mathbf{x}, t)$.

It is now quite natural to suppose: (1) that in its *own rest frame* each clock oscillates with a uniform angular frequency, which is the same for all clocks, and (2), that all clocks in the same neighbourhood are, on the average, in phase with each other. In homogeneous space there can be no reason to favour one clock over another, nor can there be a favoured direction of space (as would be implied by a non-zero average value for $\nabla\phi$ in the rest frame). We can therefore write

$$\delta\phi = \omega_0 \delta\tau \tag{17}$$

where $\delta\tau$ is the change of proper time of the clock, and where $\delta\phi$ is independent of $\delta\mathbf{x}$ in this frame.

The reason for the equality of clock phases in the rest frame for a neighbourhood can be understood more deeply as a natural consequence of the non-linearity, of the coupling of the neighbouring clocks (implied by the general non-linearity of the field equations). It is well known that two oscillators of the same natural frequency tend to come into phase with each other when there is such a coupling.[28] Of course, the relative phase will oscillate somewhat, but in the long run, and on the average, these oscillations will cancel out.

Let us now consider the problem in some fixed Lorentz frame, e.g., that of the laboratory. We then calculate the change of $\delta\phi$ $(\mathbf{x}, t)$ which would follow upon a virtual displacement $(\delta\mathbf{x}, \delta t)$. This depends only on δr. By a Lorentz transformation, we obtain

$$\delta\phi = \omega_0\delta\tau = \frac{\omega_0[\delta t - (\mathbf{v}\cdot\delta\mathbf{x})/c^2]}{\sqrt{1 - \dfrac{v^2}{c^2}}}. \tag{18}$$

If we integrate $\delta\phi$ around a closed circuit, the change of phase, $\delta\phi_c$, should then be $2n\pi$, where n is an integer. Otherwise, the clock phases would not be single-valued functions of $\mathbf{x}$ and t. We thus obtain

$$\oint \delta\phi = \omega_0 \oint \frac{(\delta t - \mathbf{v}\cdot\delta\mathbf{x}/c^2)}{\sqrt{1 - \frac{v^2}{c^2}}} = 2n\pi. \quad (19)$$

If we now suppose that each effective clock has some rest mass, m_0, and if we write for the total energy of translation of the clock, $E = m_0c^2/\sqrt{1 - (v^2/c^2)}$, and for the corresponding momentum, $\mathbf{p} = m_0\mathbf{v}/\sqrt{1 - (v^2/c^2)}$ we get

$$\oint (E\delta t - \mathbf{p}\delta\mathbf{x}) - 2n\pi \frac{m_0}{\omega_0} c^2 \quad (20)$$

If we assume that $m_0c^2/\omega_0 = \hbar$ (a universal constant for all the clocks) we obtain just the kind of quantization that we need, for circuit integrals involving the translational momentum, $\mathbf{p}$ and the coordinates of the clocks, $\mathbf{x}$ (e.g., we can set $\delta t = 0$ and Eq. (20) reduces to a special case of Eq. (16)).

We see, then, that quantization of action can, at least in this special case, arise out of certain topological conditions, implied by the need for single-valuedness of the clock phases.

The above idea provides a starting point for a deeper understanding of the meaning of the quantum conditions, but it needs to be supplemented in two ways. First, we must consider the further fluctuations in the field, associated with the non-denumerable infinity of degrees of freedom. Second, we shall have to justify the assumption that the ratio m_0c^2/ω_0 in Eq. (20) is universal for all the local clocks and equal to $\hbar$.

To begin with, we recall that each local clock of a given level exists in a certain region of space and time, which is made up of still smaller regions, and so on without limit. We shall see that we can obtain the universability of the quantum of action, h, at all levels, if we assume that each of the above *sub-regions* contains an effective clock of a similar kind, related to the other effective clocks of its level in a similar way, and that this effective clock structure continues indefinitely with the analysis of space and time into sub-regions. We stress that this is only a preliminary assumption, and that later we will show that the notion of the indefinite continuation of the above clock structure can be given up.

To treat this problem, we introduce an ordered infinity of dynamic coordinates, x_i^l, and the conjugate momenta, p_i^l. The mean position of the ith clock at the lth level of size is represented by x_i^l, and p_i^l represents the corresponding momentum. To a first approximation the quantities of each level can be treated as collective coordinates of the next lower level set of variables; but this treatment cannot in general be completely exact because *each level will to some extent be influenced directly by all the other levels, in a way that cannot fully be expressed in terms of their effects on the next lower level quantities alone.* Thus, while each level is strongly correlated to the mean behaviour of the next lower level, it has some degree of independence.

The above discussion leads us to a certain ordering of the infinity of field variables that is indicated by the nature of the problem itself. In this ordering, we consider the series of quantities, x_i^l and p_i^l, defined above as, in principle, all independent coordinates and momenta which are, however, usually connected and correlated by suitable interactions.

We can now treat this problem by means of a canonical transformation. We introduce an action function, S, which depends on all the variables, x_i^l, of the infinity of clocks within clocks. As before, we then write

$$P_k^l = \frac{\partial S}{\partial x_k^l}(x_i^l \; . \; . \; . \; x_k^l \; . \; . \; .) \qquad (21)$$

where l' represents all possible levels.

For the constants of the motion, we write

$$I_c = \sum_{k,l} \oint p_k^l \delta x_k^l = \delta S_c \qquad (22)$$

where the integrals are carried over suitable contours.

Each of these constants of the motion is now built up out of circuit integrals involving $p_i \delta x_i$, but as we saw, each one of these clocks must satisfy the phase condition $\oint p_\mu \delta_\chi{}^\mu = 2n\pi\hbar$ around any circuit. Hence the sum satisfies such a condition, which in turn must be satisified not only in real circuits actually traversed by the clocks but also in any virtual circuit that is consistent with a given set of values for the constants of the motion. Because of fluctuations coming from lower levels, there is always the possibility that any clock may move on any one of the circuits in question; and unless the constants of the motion are determined such that $\delta S_c = 2n\pi\hbar$, clocks that reach the same position after having followed different randomly fluctuating paths will not, in general, agree with each other in their phases. Thus, the agreement of the phases of all clocks that reach the same point in space and time is equivalent to the quantum condition.

The self-consistency of the above treatment can now be verified in a further analysis, which also eliminates the need to introduce the special assumption that m_0c^2/ω_0 is universally constant and equal to $\hbar$ for all clocks. Each clock is now regarded as a composite system made up of smaller clocks. Indeed, to an adequate degree of approximation, each clock phase can be treated as a collective variable associated with the *space coordinates* of the smaller clocks (which then represent the inner structure of the clock in question). Now the action variable

$$I_c = \oint_c \sum_{k,l} p_k^l \delta q_k^l$$

is canonically invariant, in the sense that it takes the same form with every set of canonical variables, and is not changed in its value by a canonical transformation. Hence, if we transformed to the collective coordinates of any given level, we would still obtain the same kind of restriction I_c to integral multiples of h, even if I_c were expressed in terms of the collective variables. Thus the collective variables of a given level will generally be subject to the same quantum restriction as those satisfied by the original variables of that level. In order that it be consistent for variables of a given level to be essentially equal to collective variables for the next lower level, it is sufficient that the variables of all levels be quantized in terms of the same unit of action, h. In this way, an overall consistent ordering of the non-denumerable infinity of variables becomes possible.

Each clock will then have a quantized value for the action variable, I_c, associated with its *inner motion* (i.e., of its phase changes). This inner motion was, however, assumed to be effectively that of a harmonic oscillator. Therefore, according to a well-known classical result, the inner energy is $E = J\omega_0/2\pi$; and since $J = Sh$, where S may be any integer, we obtain $E_0 = S\omega_0\hbar$.

Now E_0 is also the rest energy of the clock, so that $E_0 = m_0c^2$. hence we obtain

$$\frac{m_0 c^2}{\omega_0} = S\hbar. \tag{23}$$

This gives us, from Eq. (20),

$$\oint (E\delta t - \mathbf{p}\delta \mathbf{x}) = 2\pi \frac{m_0 c^2}{\omega_0} n = nSh = nh; \tag{24}$$

and since, in general, S takes on arbitrary integral values, it is also an arbitrary integer. In this way, we eliminate the need for assuming separately, that $m_0 c^2/\omega_0$ is a universal constant, equal to $\hbar$.

To finish this stage of the development of the theory, we must show that the model discussed above leads to a fluctuation in the phase space of the variables of a given level, in accordance with that implied by Heisenberg's principle. In other words, the quantum of action, h, must also be shown to yield a correct estimate of the limitation on the degree of self-determination of the quantities of any level.

To prove the above conjecture, we must note that each variable fluctuates because it depends on the lower-level quantities (of which it is a collective coordinate). The lower-level quantities can change their action variables only by discrete multiples of h. It is therefore not implausible that the domain of fluctuation of a given variable would be closely related to the size of the possible discrete changes in its constituent lower-level variables.

We shall prove the theorem stated above for the special case that all the degrees of freedom can be represented as coupled harmonic oscillators. This is a simplification of the real problem (which is non-linear). The real motions will consist of small systematic perturbations on top of an infinitely turbulent background. These systematic perturbations can be treated as collective coordinates, representing the overall behaviour of the constituent local clocks of a given level. In general, such a collective motion will take the form of a wavelike oscillation, which to a certain degree of approximation undergoes simple harmonic motion. Let us represent the action variables and angle-variables of the nth harmonic oscillator by J_n and ϕ_n respectively. To the extent that the linear approximation is correct, J_n will be a constant of the motion, and ϕ_n will increase linearly with time according to the equation $\phi_n = \omega_n t + \phi_{0n}$, where ω_n is the angular frequency of the nth oscillator. J_n and ϕ_n will be related to the clock variables by a canonical transformation, such as (12). Because the generalized Bohr-Sommerfeld correlation (16) is invariant to a canonical transformation, it follows that $J_n = Sh$, where S is an integer. Moreover, the coordinates and momenta of these oscillators can be written as[29]

$$p_n = 2\sqrt{J_n}\cos\phi_n, \quad q_n = 2\sqrt{J_n}\sin\phi_n.$$

We now consider a higher level canonical set of variables, a specific pair of which we denote by Q_i^l and π_i^l. In principle, these would be determined by the totality of all the other levels. To be sure, the next lower level will be the *main* one that enters into this determination; nevertheless, the others will still have *some* effect. Hence in accordance with our earlier discussions, we must regard π_i^l and Q_i^l as being, in principle, independent of any *given* set of lower-level variables, including, of course, those of the next lower level.

To the extent that the linear approximation is valid, we can write[30]

$$Q_i^l = \sum_n \alpha_{in} p_n = 2\sum_n \alpha_{in}\sqrt{J_n}\cos\phi_n$$

$$\pi_i^l = \sum_n \beta_{in} q_n = 2\sum_n \beta_{in}\sqrt{J_n}\sin\phi_n \tag{25}$$

where α_{in} and β_{in} are constant coefficients, and where, as we recall, n is assumed to cover *all* levels other than l.

In order that it be consistent to suppose that Q_i^l and π_i^l are canonical conjugates, it is necessary that their Poisson bracket be unity or that

$$\sum_n \left(\frac{\partial \pi_i^l}{\partial J_n}\frac{\partial Q_i^l}{\partial \phi_n} - \frac{\partial \pi_i^l}{\partial \phi_n}\frac{\partial Q_i^l}{\partial J_n} \right) = 1.$$

With the aid of Eq. (25), this becomes

$$\Sigma \alpha_n \beta_n = 1. \tag{26}$$

Eq. (25) implies a very complex motion for Q_i^l and π_i^l, for in a typical system of coupled oscillators the ω_n are in general all different and are not integral multiples of each other (except for possible sets of measure zero). Thus, the motion will be a 'space-filling' (quasi-ergodic) curve in phase space, being a generalization of the two-dimensional Lissajou figures for perpendicular harmonic oscillators, with periods that are not rational multiples of each other.

During a time interval, τ, which is fairly long compared with the periods, $2\pi/\omega_n$, of the lower-level oscillators, the trajectory of Q_i^l and π_i^l in the phase space will, in essence, fill a certain region, even though the orbit is definite at all times. We shall now calculate the mean fluctuation of Q_i^l and π_i^l in this region by taking averages over the time, τ. Noting that $\overline{Q_i^l} = \overline{\pi_i^l} = 0$ for such averages, we have for these fluctuations,

$$(\Delta Q_i^l)^2 = 4 \sum_{mn} \alpha_m \alpha_n \sqrt{J_m J_n \cos\phi_m \cos\phi_n} = 2 \sum_m (\alpha_m)^2 J_m \tag{27}$$

$$(\Delta \pi_i^l)^2 = 4 \sum_{mn} \beta_m \beta_n \sqrt{J_m J_n \sin\phi_m \sin\phi_n} = 2 \sum_n (\beta_n)^2 J_n \tag{28}$$

where we have used the result that $\cos \delta_m \cos \delta_n = \sin \delta_m \sin \delta_n = 0$ for $m \neq n$ (except for the set of zero measure, mentioned above, in which ω_m and ω_n are rational multiples of each other).

We now suppose that all the oscillators are in their lowest states (with $J = h$) except for a set of zero measure. This set represents a denumerable number of excitations relative to the 'vacuum' state. Because of their small number, these make a negligible contribution to $(\Delta Q_i^l)^2$ and $(\Delta \pi_i^l)^2$.

We therefore set $J_n = h$ in Eq. (28) and obtain

$$(\Delta Q_i^l)^2 = 2 \sum_m (\alpha_m)^2 h; \; (\Delta \pi_i^l)^2 = 2 \sum_n (\beta_n)^2 h.$$

We then use the Schwarz inequality

$$\sum_{mn} (\alpha_m)^2 (\beta_n)^2 \geqslant | \sum_m \alpha_m \beta_m |^2. \tag{29}$$

Combining the above with Eqs. (26), (27) and (28), we obtain

$$(\Delta\pi_i^l)^2\,(\Delta Q_i^l)^2 \geqslant 4h^2. \tag{30}$$

The above relations are, in essence, those of Heisenberg. $\Delta\pi_i^l$ and ΔQ_i^l will effectively represent limitations on the degree of self-determination of the lth level, because all quantities of this level will evidently have to be averaged over periods of time long compared with $2\pi/\omega_n$. Thus, we have deduced Heisenberg's principle from the assumption of the quantum of action.

We note that Eq. (30) has already been obtained in section 10 in a very different way – by assuming simple random field fluctuations resembling those of particles undergoing Brownian motion. Hence, an infinity of lower-level variables satisfying the conditions that J_n is discrete and equal to the same constant, h, for all the variables, will yield a long-run pattern of motions that reproduces certain essential features of a random Brownian-type fluctuation.

We have thus completed our task of proposing a general physical model that explains the quantization rules along with the Heisenberg indeterminacy relations. But now, it can easily be seen that our basic physical model, involving an infinity of clocks within clocks, leaves room for fundamental changes, which would go outside the scope of the current quantum theory. To illustrate these possibilities, suppose that such a structure were to continue only for some characteristic time, τ_0, after which it would cease to exist and would be replaced by another kind of structure. Then, in process that involve times much greater than τ_0, the clocks will still be restricted in essentially the same ways as before, since their motions would not significantly be changed by the deeper substructure. Nevertheless, in processes involving times shorter than τ_0, there will be no reasons for such restrictions to apply, since the structure is no longer the same. In this way, we see how J_n will be restricted to discrete values in certain levels, while they are not necessarily restricted in this way in other levels.

For levels in which J_n are not restricted to being multiples of h, Eq. (30) for the fluctuation of π_i^l and Q_i^l need no longer apply. In place of h, there will appear a quantity, J_m, the *mean* action associated with the levels in question. In addition, averages of $(\cos\phi_m \cos\phi_n)$ may cease to be negligible, because the times are too short. Thus, there is room for every conceivable, kind of change in the rules for determining J_n and in those determining the magnitudes of fluctuation associated with a given level. Nevertheless, in the quantum levels the usual rules will be valid to a very high degree of approximation.

13 DISCUSSION OF EXPERIMENTS TO PROBE SUB-QUANTUM LEVEL

We are now ready to discuss, at least in general terms, the conditions under which it might be possible to test for a sub-quantum level experimentally, and in this way to complete our answers to the criticisms of the suggestion of hidden variables made by Heisenberg and Bohr.

We first recall that the proof of Heisenberg's relations, concerning the maximum possible accuracy of measurement of canonically conjugate variables, made use of the implicit assumption that measurements must involve only processes satisfying the general laws of the current quantum theory. Thus, in the well-known example of the gamma-ray microscope, he supposed that the position of an electron was to be measured by scattering a gamma ray from the particle in question into a lens and on to a photographic plate. This scattering is essentially a case of the Compton effect; and the proof of Heisenberg's principle depended essentially on the assumption that the Compton effect satisfies the laws of the quantum theory (i.e., conservation of energy and momentum in an 'indivisible' scattering process, wavelike character of the scattered quantum as it goes through the lens, and incomplete determinism of the particle-like spot on the photographic plate). More generally, any such proof must be based on the assumption that at every stage the process of measurement will satisfy the laws of the quantum theory. Thus to suppose that Heisenberg's principle has a universal validity is, in the last analysis, the same as to suppose that the general laws of the quantum theory are universally valid. But this supposition is now expressed in terms of the *external relations* of the particle to a measuring apparatus, and not in terms of the inner characteristics of the particle itself.

In our point of view, Heisenberg's principle should not be regarded as *primarily* an external relation, expressing the impossibility of making measurements of unlimited precision in the quantum domain. Rather, it should be regarded as basically an expression of the incomplete degree of *self-determination* characteristic of all entities that can be defined in the quantum-mechanical level. It follows that if we measure such entities, we will also use processes taking place in the quantum-mechanical level, so that the process of measurement will have the same limits on its degree of self-determination as every other process in this level. It is rather as if we were measuring Brownian motion with microscopes that were subject to the same degree of random fluctuation as that of the systems that we were trying to observe.

As we saw in sections 10 and 12, however, it is possible and indeed rather plausible to suppose that sub-quantum-mechanical processes involving very small intervals of time and space will not be subject to the same limitations of their degree of self-determination as those of quantum-mechanical processes. Of course, these sub-quantum processes will very probably involve basically new kinds of entities as different from electrons, protons, etc., as the latter are from macroscopic systems. Hence, entirely new methods would have to be developed to observe them (as new methods also had to be developed to observe atoms, electrons, neutrons, etc.). These methods will depend on using interactions involving sub-quantum laws. In other words, just as the 'gamma-ray microscope' was based on the existence of the Compton effect, a 'sub-quantum microscope' would be based on new effects, not limited in their degree of self-determination by the laws of the quantum theory. These effects would then make possible a correlation between an observable large-scale event and the state of some sub-quantum variable, that is more accurate than is permitted in Heisenberg's relations.

Of course, one does not expect, in the way described above, to actually determine all the sub-quantum variables and thus to predict the future in full detail. Rather, one aims only by a few crucial experiments to show that the sub-quantum level is there, to investigate its laws, and to use these laws to explain and predict the properties of higher-level systems in more detail, and with greater precision than the current quantum theory does.

To treat this question in more detail, we now recall a conclusion of the previous section, i.e. that if in lower levels the action variable should be divisible in units smaller than h, then the limits on the degree of self-determination of these lower levels can be less severe than those given by Heisenberg's relations. Thus, there may well be relatively divisible and self-determined processes going on at lower levels. But how can we observe them on our level?

To answer the above question, we refer to Eq. (25), which indicates in typical case how the variables of a given level depend to some extent on *all* the lower-level variables. Thus if π_i^l and Q_i^l

represent the classical level, then they would, in general, be determined *mainly* by the p_i^l and q_i^l of the quantum level; but there would be *some* effects due to sub-quantum levels. Usually these would be quite small. However, in special cases (e.g., with special arrangements of apparatus) the π_i^l and Q_i^l might depend significantly on the p_i^l and q_i^l of a sub-quantum level. Of course, this would mean the coupling of some new kind of sub-quantum process (as yet unknown, but perhaps to be discovered later) to the observable large-scale classical phenomena. Such a process would presumably involve high frequencies and therefore high energies, but in a new way.

Even when the effects of the sub-quantum level on π_i^l and Q_i^l are small, they are not identically zero. Thus, room is created for testing for such effects by doing old kinds of experiments with extremely high precision. For example, the relation $J_n = nh$ was obtained in Eq. (24) only if the quantum of action was supposed to be universally equal to h (at all levels). Sub-quantum deviations from this rule would therefore be reflected in the classical level as a minute error in the relation $E = nh\nu$ for a harmonic oscillator. In this connection, recall that in classical theory there is no special relation between energy and frequency at all. This situation may to some extent be restored in the sub-quantum domain. As a result, one would discover a small fluctuation in the relation between E_n and $nh\nu$. For example, one would have

$$E_n = nh\nu + \epsilon$$

where ϵ is a very small randomly fluctuating quantity (which gets larger and larger as we go to higher and higher frequencies). To test for such a fluctuation, one could perform an experiment in which the frequency of a light beam was observed to an accuracy, $\nabla\nu$. If the observed energy fluctuated by more than $\hbar\,\nabla\nu$, and if no source could be found for the fluctuation in the quantum level, this experiment could be taken as an indication of sub-quantum fluctuations.

With this discussion, we complete our answer to the criticisms of Bohr and Heisenberg, who argue that a deeper level of hidden variables in which the quantum of action was divisible could never be revealed in any experimental phenomena. This also means that there are no valid arguments justifying the conclusion of Bohr that the concept of the detailed behaviour of matter as a unique and self-determining process must be restricted to the classical level only (where one can observe fairly directly the behaviour of the large-scale phenomena). Indeed we are also able to apply such notions in a sub-quantum level, whose relations with the classical level are relatively indirect, and yet capable, in principle, of revealing the existence and the properties of the lower level through its effects to the classical level.

Finally, we consider the paradox of Einstein, Rosen and Podolsky. As we saw in section 4, we can easily explain the peculiar quantum-mechanical correlations of distant systems by supposing hidden interactions between such systems, carried in the sub-quantum level. With an infinity of fluctuating field variables in this lower level, there are ample motions going on that might explain such a correlation. The only real difficulty is to explain how the correlations are maintained if, while the two systems are still flying apart, we suddenly change the variable that is going to be measured by changing the measuring apparatus for one of the systems. How, then, does the far-away system receive instantaneously a 'signal' showing that a new variable is going to be measured, so that it will respond accordingly?

To answer this question, we first note that the characteristic quantum-mechanical correlations have been observed experimentally with distant systems only when the various pieces of observing apparatus have been standing around so long that there has been plenty of opportunity for them to come to equilibrium with the original system through sub-quantum-mechanical interactions.[31] For example, in the case of the molecule described in section 4, there would be time for many impulses to travel back and forth between the molecule and the spin-measuring devices, even before the molecule disintegrated. Thus, the actions of the molecule could be 'triggered' by signals from the apparatus, so that it would emit atoms with spins already properly lined up for the apparatus that was going to measure them.

In order to test the essential point here, one would have to use measuring systems that were changed rapidly compared with the time needed for a signal to go from the apparatus to the observed system and vice versa. What would really happen if this were done is not yet known. It is possible that the experiments would disclose a failure of the typical quantum-mechanical correlations. If this were to happen, it would be a proof that the basic principles of the quantum are breaking down here, for the quantum theory could not explain such a behaviour, while a sub-quantum theory could quite easily explain it as an effect of the failure of sub-quantum connections to relate the system systems rapidly enough to guarantee correlations when the apparatus was changed very suddenly.

On the other hand, if the predicted quantum-mechanical correlations are still found in such a measurement, this is no proof that a sub-quantum level does not exist, for even the mechanical device that suddenly changes the observing apparatus must have sub-quantum connections with all parts of the system, and through these a 'signal' might still be transmitted to the molecule that a certain observable was eventually going to be measured. Of course, we would expect that at some level of complexity of the apparatus, the sub-quantum connections would cease to be able to do this. Nevertheless, in the absence of a more detailed sub-quantum-mechanical theory, where this will happen cannot be known a priori. In any case, the results of such an experiment would surely be very interesting.

14 CONCLUSION

In conclusion, we have carried the theory far enough to show that we can explain the essential features of the quantum mechanics in terms of a sub-quantum-mechanical level involving hidden variables. Such a theory is capable of having a new experimental content, especially in connection with the domain of very short distances and very high energies, where there are new phenomena not very well treated in terms of present theories (and also in connection with the experimental verification of certain features of the correlations of distant systems). Moreover, we have seen that this type of theory opens up new possibilities for elimination divergence in present theories, also associated with the domain of short distances and high energies. (For example, as shown in section 10, the breakdown of Heisenberg's principle for short time could eliminate the infinite effects of quantum fluctuations.)

Of course, the theory as developed here is far from complete. It is necessary at least to show how one obtains the many-body Dirac equation for fermions, and the usual wave equations for bosons. On these problems much progress has been made but there is no space to enter into a discussion of them here. In addition, further progress is being made on the systematic treatment of the new kinds of particles (mesons, hyperons, etc.) in terms of our scheme. All of this will be published later and elsewhere.

Nevertheless, even in its present incomplete form, the theory does answer the basic criticisms of those who regarded such a theory as impossible, or who felt that it could never concern itself with any real experimental problems. At the very least, it does seem to have promise of being able to throw some light on a number of such experimental problems, as well as on those arising in connection with the lack of internal consistency of the current theory.

For the reasons described above, it seems that some consideration of theories involving hidden variables is at present needed to help us to avoid dogmatic preconceptions. Such preconceptions not only restrict our thinking in an unjustifiable way but also similarly restrict the kinds of experiments that we are likely to perform (since a considerable fraction of all experiments is, after all, designed to answer questions raised in some theory). Of course, it would be equally dogmatic to insist that the usual interpretation has already exhausted all of its possible usefulness for these problems. What is necessary at the present time is that many avenues of research be pursued, since it is not possible to know beforehand which is the right one. In addition, the demonstration of the possibility of theories of hidden variables may serve in a more general philosophical sense to remind us of the unreliability of conclusions based on the assumption of the complete universality of certain features of a given theory, however general their domain of validity seems to be.

5

Quantum theory as an indication of a new order in physics

Part A: The development of new orders as shown through the history of physics

1 INTRODUCTION

Revolutionary changes in physics have always involved the perception of new order and attention to the development of new ways of using language that are appropriate to the communication of such order.

We shall start this chapter with a discussion of certain features of the history of the development of physics that can help give some insight into what is meant by perception and communication of a new order. We shall then go on in the next chapter to present our suggestions regarding the new order that is indicated by the consideration of the quantum theory.

In ancient times, there was only a vague qualitative notion of order in nature. With the development of mathematics, notably arithmetic and geometry, the possibility arose for defining forms and ratios more precisely, so that, for example, one could describe the detailed orbits of planets etc. However, such detailed mathematical descriptions of the motions of the planets and other heavenly bodies implied certain general notions of order. Thus, the Ancient Greeks thought that the Earth was at the centre of the universe, and that surrounding the Earth were spheres, which approached the ideal perfection of celestial matter as one got further and further away from the Earth. The perfection of celestial matter was supposed to be revealed in circular orbits, which were regarded as the most perfect of all geometrical figures, while the imperfection of earthly matter was thought to be shown in its very complicated and apparently arbitrary movements. Thus, the universe was both perceived and discussed in terms of a certain overall order; i.e., the order of degrees of perfection, which corresponded to the order of distance from the centre of the Earth.

Physics as a whole was understood in terms of notions of order closely related to those described above. Thus, Aristotle compared the universe to a living organism, in which each part had its proper place and function, so that all worked together to make a single whole. Within this whole, an object could move only if there was a force acting on it. Force was thus thought of as a *cause* of motion. So the order of movement was determined by the order of causes, which in turn depended on the place and function of each part in the whole.

The general way of perceiving and communicating order in physics was, of course, not at all in contradiction with common experience (in which, for example, movement is possible as a rule only when there is a force which overcomes friction). To be sure, when more detailed observations were made on the planets, it was found that their orbits are not actually perfect circles, but this fact was accommodated within the prevailing notions of order by considering the orbits of planets as a superposition of *epicycles*, i.e., circles within circles. Thus, one sees an example of the remarkable capacity for *adaptation* within a given notion of order, adaptation that enables one to go on perceiving and talking in terms of essentially fixed notions of this kind in spite of factual evidence that might at first sight seem to necessitate a thoroughgoing change in such notions. With the aid of such adaptations, men could for thousands of years look at the night sky and see epicycles there, almost independently of the detailed content of their observations.

It seems clear, then, that a basic notion of order, such as was expressed in terms of epicycles, could never be decisively contradicted, because it could always be adjusted to fit the observed facts. But eventually, a new spirit arose in scientific research, which led to the questioning of the *relevance* of the old order, notably by Copernicus, Kepler, and Galileo. What emerged from such questioning was in essence the proposal that the difference between earthly and celestial matter is not actually very significant. Rather, it was suggested that a key difference is between the motion of matter in empty space and its motion in a viscous medium. The basic laws of physics should then refer to the motion of matter in empty space, rather than to its motion in a viscous medium. Thus, Aristotle was right to say that matter as commonly experienced moved only under the action of a force, but he was wrong in supposing that this common experience was relevant to the fundamental laws of physics. From this it followed that the key difference between celestial and earthly matter was not in its degree of perfection but rather in that celestial matter generally moves without friction in a vacuum, whereas terrestrial matter moves with friction in a viscous medium.

Evidently, such notions were not generally compatible with the idea that the universe is to be regarded as a single living organism. Rather, in a fundamental description, the universe now had to be regarded as analysable into separately existing parts or objects (e.g. planets, atoms, etc.) each moving in a void or vacuum. These parts could work together in interaction more or less as do the parts of a machine, but could not grow, develop, and function in response to ends determined by an 'organism as a whole'. The basic order for description of movement of the parts of this 'machine' was taken to be that of successive positions of each constituent object at successive moments of time. Thus, a new order became relevant, and a new usage of language had to be developed for the description of this new order.

In the development of new ways of using language, the Cartesian coordinates played a key part. Indeed, the very word 'coordinate' implies a function of *ordering*. This ordering is achieved with the aid of a grid. This is constituted of three perpendicular sets of uniformly spaced lines. Each set of lines is evidently an order (similar to the order of the integers). A given curve is then determined by a *coordination* among the X, the Y and the Z orders.

Coordinates are evidently not to be regarded as natural objects. Rather, they are merely convenient forms of description set up by us. As such, they have a great deal of arbitrariness or conventionality (e.g., in orientation, scale, orthogonality, etc., of coordinate frames). Despite this kind of arbitrariness, however, it is possible, as is now well known, to have a non-arbitrary general law expressed in terms of coordinates. This is possible if the law takes the form of a relationship that remains *invariant* under changes in the arbitrary features of the descriptive order.

To use coordinates is in effect to order our attention in a way that is appropriate to the mechanical view of the universe, and thus similarly to order our perception and our thinking. It is clear, for example, that though Aristotle very probably would have understood the meaning of coordinates, he would have found them of little or no significance for his aim of understanding the universe as an organism. But once men were ready to conceive of the universe as a machine, they would naturally tend to take the order of coordinates as a universally relevant one, valid for all basic descriptions in physics.

Within this new Cartesian order of perception and thinking that had grown up after the Renaissance, Newton was able to discover a very general law. It may be stated thus: 'As with the order of movement in the fall of an apple, so with that of the Moon, and so with *all*.' This was a new perception of law, i.e., universal harmony in the order of nature, as described in detail through the use of coordinates. Such perception is a flash of very penetrating insight, which is basically *poetic*. Indeed, the root of the word 'poetry' is the Greek 'poiein', meaning 'to make' or 'to create'. Thus, in its most original aspects, science takes on a quality of poetic communication of creative perception of new order.

A somewhat more 'prosaic' way of putting Newton's insight is to write $A:B::C:D$. That is to say: 'As the successive positions A, B of the apple are related, so are the successive positions C, D of the Moon.' This constitutes a generalized notion of what may be called *ratio*. Here, we take ratio in its broadest meaning (e.g., in its original Latin sense) which includes all of *reason*. Science thus aims to discover universal ratio or reason, which includes not only numerical ratio or proportion ($A/B = C/D$), but also general qualitative similarity.

Rational law is not restricted to an expression of *causality*. Evidently, reason, in the sense that is meant here, goes far beyond causality, which latter is a special case of reason. Indeed, the basic form of causality is: 'I do a certain action X and cause something to happen.' A causal law then takes the form: 'As with such causal actions of mine, so with certain processes that can be observed in nature.' Thus, a causal law provides a certain *limited kind* of reason. But, more generally, a rational explanation takes the form: 'As things are related in a certain idea or concept, so they are related in fact.'

It is clear from the preceding disucssion that in finding a new structure of reason or rationality, it is crucial *first* to discern relevant differences. To try to find a rational connection between irrelevant differences leads to arbitrariness, confusion, and general sterility (e.g., as with epicycles). So we have to be ready to drop our assumptions as to what are the relevant differences, though this has often seemed to be very difficult to do, because we tend to give such high psychological value to familiar ideas.

2 WHAT IS ORDER?

Thus far, the term order has been used in a number of contexts that are more or less known to everyone, so that its meaning can be seen fairly clearly from its usage. The notion of order, however, is evidently relevant in much broader contexts. Thus, we do not restrict order to some regular arrangement of objects or forms in lines or in rows (e.g., as with grids). Rather, we can consider much more general orders, such as the order of growth of a living being, the order of evolution of living species, the order of society, the order of a musical composition, the order of painting, the order which constitutes the meaning of communication, etc. If we wish to inquire into such broader contexts, the notions of order to which we have referred earlier in this chapter will evidently no longer be adequate. We are therefore led to the general question: 'What is order?'

The notion of order is so vast and immense in its implications, however, that it cannot be defined in words. Indeed, the best we can do with order is to try to 'point to it' tacitly and by implication, in as wide as possible a range of contexts in which this notion is relevant. We all know order implicitly, and such 'pointing' can perhaps communicate a general and overall meaning of order without the need for a precise verbal definition.

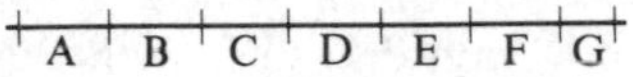

A B C D E F G

Figure 5.1

To begin to understand order in such a general sense, we may first recall that in the development of classical physics the perception of a new order was seen to involve the discrimination of new relevant differences (positions of objects at successive moments of time) along with new similarities that are to be found in the differences (similarity of 'ratios' in these differences). It is being suggested here that this is the seed or nucleus of a very general way of perceiving order, i.e., *to give attention to similar differences and different similarities*.[1]

Let us illustrate these notions in terms of a geometric curve. To simplify the example, we shall approximate the curve by a series of straight-line segments of equal length. We begin with a straight line. As shown in Figure 5.1, the segments in a straight line all have the same direction, so that their only difference is in the position. The difference between segment A and segment B is thus a space displacement which is similar to the difference between B and C, and so on. We may therefore write

$$A{:}B{::}B{:}C{::}C{:}D{::}D{:}E.$$

This expression of 'ratio' or 'reason' may be said to define a curve of *first class*, i.e., a curve having only one independent difference.

Next, we consider a circle, as illustrated in Figure 5.2. Here, the difference between A and B is in direction as well as in position. Thus, we have a curve with two independent differences – which is therefore one of *second class*. However, we still have a single 'ratio' in the differences, $A{:}B{::}B{:}C$.

Now we come to a helix. Here, the angle between lines can turn in a third dimension. Thus, we have a curve of *third class*. It, too, is determined by a single ratio, $A{:}B{::}B{:}C$.

Thus far we have considered various *kinds* of similarities in the differences, to obtain curves of first, second, third classes, etc. However, in each curve, the similarity (or ratio) between successive steps remains invariant. Now we can call attention to curves in which *this similarity is different* as we go along the curve. In this way, we are led to consider not only *similar differences* but also *different similarities of the differences*.

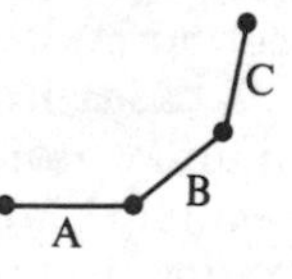

Figure 5.2

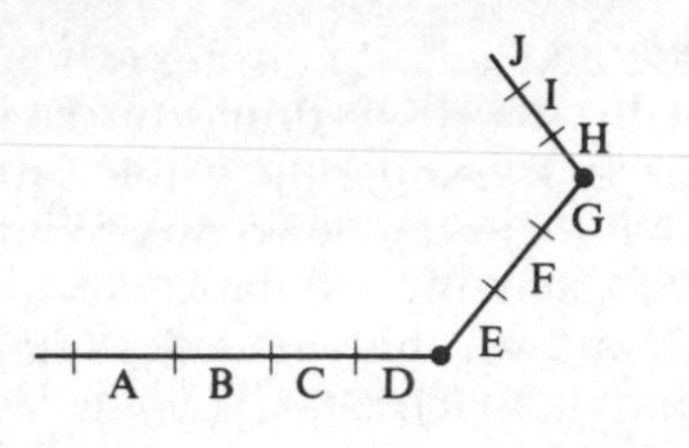

Figure 5.3

We can illustrate this notion by means of a curve which is a chain of straight lines in different directions (see Figure 5.3). On the first line $(ABCD)$, we can write

$$A:B^{S_1}::B:C.$$

The symbol S_1 stands for 'the first kind of similarity', i.e., in direction along the line $(ABCD)$. Then we write for the lines (EFG) and (HIJ)

$$E:F^{S_2}::F:G \quad \text{and} \quad H:I^{S_3}::I:J$$

where S_2 stands for 'the similarity of the second kind' and S_3 for 'the similarity of the third kind'.

We can now consider the difference of successive similarities $(S_1, S_2, S_3, \ldots)$ as a *second degree of difference*. From this, we can develop a *second degree of similarity in these differences*. $S_1:S_2::S_2:S_3$.

By thus introducing what is in effect the beginning of a hierarchy of similarities and differences, we can go on to curves of arbitrarily high degrees of order. As the degrees become indefinitely high, we are able to describe what have commonly been called 'random' curves – such as those encountered in Brownian motion. This kind of curve is not determined by any finite number of steps. Nevertheless, it would not be appropriate to call it 'disordered', i.e., *having no order whatsoever*. Rather, it has a certain kind of order, which is of an indefinitely high degree.

In this way, we are led to make an important change in the general language of description. We no longer use the term 'disorder' but instead we distinguish between different degrees of order (so that, for example, there is an unbroken gradation of curves, beginning with those of first degree, and going on step by step to those that have generally been called 'random').

It is important to add here that order is not to be identified with *predictability*. Predictability is a property of a special kind of order such that a few steps determine the whole order (i.e., as in curves of low degree) – but there can be complex and subtle orders which are not in essence related to predictability (e.g. a good painting is highly ordered, and yet this order does not permit one part to be predicted from another).

3 MEASURE

In developing the notion of an order of high degree, we have tacitly brought in the idea that each sub-order has a *limit*. Thus, in Figure 5.4 the order of the line *ABC* reaches its limit at the end of the segment *C*. Beyond this limit is another order, *EFG*, and so on. So, the description of a hierarchic order of high degree generally involves the notion of limit.

It is significant to note here that in ancient times the most basic meaning of the word 'measure' was 'limit' or 'boundary'. In this sense of the word, each thing could be said to have its appropriate measure. For example, it was thought that when human behaviour went beyond its proper bounds (or measure) the result would have to be tragedy (as was brought out very forcefully in Greek dramas). Measure was indeed considered to be as essential to the understanding of the good. Thus, the origin of the word 'medicine' is the Latin 'mederi', which means 'to cure' and which was derived from a root meaning 'measure'. This implied that to be healthy was to have everything in a right measure, in body and mind. Similarly, wisdom was equated with *moderation* and *modesty* (whose common root is also derived from measure), thus suggesting that the wise man is the one who keeps everything in the right measure.

To illustrate this meaning of the word 'measure' in physics, one could say that 'the measure of water' is between 0° and 100°C. In other words, measure primarily gives the limits of qualities or of orders of movement and behaviour.

Of course, measure has to be *specified* through proportion or ratio, but, in terms of the ancient notion, this specification is understood as secondary in significance to the boundary or limit which is thus specified; and here one can add that in general this specification need not even be in terms of quantitative proportion, but rather can be in terms of qualitative reason (e.g., in a drama the proper measure of human behaviour is specified in qualitative terms rather than by means of numerical ratios).

In the modern usage of the word 'measure', the aspect of quantitative proportion or numerical ratio tends to be emphasized much more heavily than it was in ancient times. Yet even here the notion of boundary or limit is still present, though in the background. Thus, to set up a *scale* (e.g., of length) one must establish divisions which are in effect *limits* or *boundaries* of ordered segments.

By giving attention in this way to older meanings of words along with their current meanings, one can obtain a certain insight into the full significance of a general notion, such as measure, which is not provided by considering only more specialized modern meanings that have been developed in various forms of scientific, mathematical and philosophical analysis.

4 STRUCTURE AS A DEVELOPMENT FROM ORDER AND MEASURE

If we consider measure in the broad sense indicated above, we can see how this notion works together with that of order. Thus, as shown in Figure 5.4 any linear order within a triangle (such as the line *FG*) is bounded (i.e., measured) by the lines *AB*, *BC*, and *CA*. Each of these lines is itself an order of segments, which is limited (i.e., measured) by the other lines. The shape of the triangle is then described in terms of certain proportions between the sides (relative lengths).

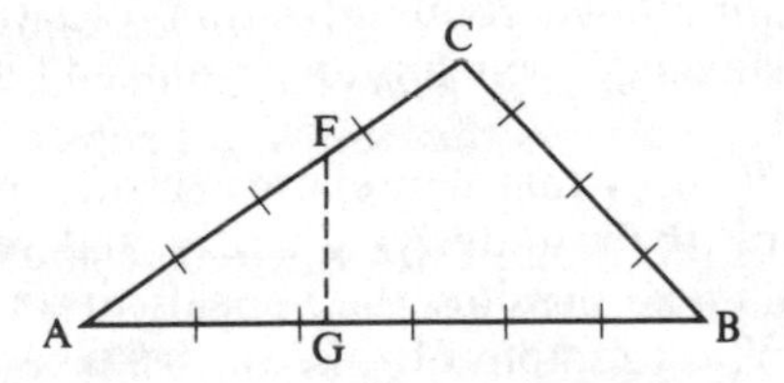

Figure 5.4

The consideration of the working together of order and measure in ever-broader and more complex contexts leads to the notion of *structure*. As the Latin root 'struere' indicates, the essential meaning of the notion of structure is to build, to grow, to evolve. This word is now treated as a noun, but the Latin suffix 'ura' originally meant 'the action of doing something'. To emphasize that we are not referring mainly to a 'finished product' or to an ultimate result, we may introduce a new verb, *to structate*, meaning 'to create and dissolve what are now called structures'.

Structation is evidently to be described and understood through order and measure. For example, consider the structation (construction) of a house. The bricks are arranged in an order and in a measure (i.e., within limits) to make walls. The walls are similarly ordered and measured to make rooms, the rooms to make a house, the houses to make streets, the streets to make cities, etc.

Structation thus implies a *harmoniously organized* totality of order and measures, which is both *hierarchic* (i.e., built on many levels) and *extensive* (i.e., 'spreading out' on each level). The Greek root of the word 'organize' is 'ergon' which is based on a verb meaning 'to work'. So one may think of all aspects of a structure as 'working together' in a coherent way.

Evidently, this principle of structure is universal. For example, living beings are in a continual movement of growth and evolution of structure, which is highly organized (e.g., molecules work together to make cells, cells work together to make organs, organs to make the individual living being, individual living beings a society, etc). Similarly, in physics, we describe matter as constituted of moving particles (e.g. atoms) which work together to make solids, liquids, or gaseous structures, which similarly make larger structures, going on up to planets, stars, galaxies, galaxies of galaxies, etc. Here, it is important to emphasize the *essentially dynamic* nature of structation, in inanimate nature, in living beings, in society, in human communication, etc. (e.g., consider the structure of a language, which is an organized totality of ever-flowing movement).

The kinds of structures that can evolve, grow, or be built are evidently limited by their underlying order and measure. New order and measure make possible the consideration of new kinds of structure. A simple example of this can be taken from music. Here, the structures that can be worked with depend on the order of the notes and on certain measures (scale, rhythm, time, etc.). New orders and measures evidently make possible the creation of new structures in music. In this chapter, we are inquiring into how new orders and measures in physics may similarly make possible the consideration of new structures in physics.

5 ORDER, MEASURE AND STRUCTURE IN CLASSICAL PHYSICS

As has already been indicated in general terms, classical physics implies a certain basic descriptive order and measure. This may be characterized as the use of certain Cartesian coordinates and by the notion of universal and absolute order of time, independent of that of space. This further implies the absolute character of what may be called *Euclidean* order and measure (i.e., that characteristic of Euclidean geometry). With this order and measure, certain structures are possible. In esssence, these are based on the quasi-rigid body, considered as a constituent element. The general characteristic of classical structure is just the analysability of everything into separate parts, which are either small, quasi-rigid bodies, or their ultimate idealization as extensionless particles. As pointed out earlier, these parts are considered to be working together in interaction (as in a machine).

The laws of physics, then, express the reason or ratio in the movements of all the parts, in the sense that the law relates the movement of each part to the configuration of all the other parts. This law is deterministic in form, in that the only contingent features of a system are the initial positions and velocities of all its parts. It is also *causal*, in that any external disturbance can be treated as a *cause*, which produces a specifiable *effect* that can in principle be propagated to every part of the system.

With the discovery of Brownian motion, one obtained phenomena that *at first sight* seemed to call the whole classical scheme of order and measure into question, for movements were discovered which were what have been called here 'order of unlimited degree', not determined by a few steps (e.g., initial positions and velocities). However, this was explained by supposing that whenever we have Brownian motion this is due to very complex impacts from smaller particles or from randomly fluctuating fields. It is then further supposed that when these additional particles and fields are taken into account, the total law will be deterministic. In this way, classical notions of order and measure can be *adapted*, so as to *accommodate* Brownian motion, which would at least on the face of the matter seem to require description in terms of a very different order and measure.

The possibility of such adaptation evidently depends, however, on an assumption. Indeed, even if we can trace *some* kinds of Brownian motion (e.g. of smoke particles) back to impacts of smaller particles (atoms), this does not prove that the laws are ultimately of the classical, deterministic kind – for it is always possible to suppose that basically all movements are to be described *from the very outset* as Brownian motion (so that the apparently continuous orbits of large objects such as planets would only be approximations to an actually Brownian type of path). Indeed, mathematicians (notably Wiener) have both implicitly and explicitly worked in terms of Brownian motion as a basic description[2] (not explained as a result of impacts of finer particles). Such an idea would in effect bring in a new kind of order and measure. If it were pursued seriously, this would imply a change of possible structures that would perhaps be as great as that implied by the change from Ptolemaic epicycles to Newtonian equations of motion. Actually, this line was not seriously pursued in classical physics. Nevertheless, as we shall see later, it may be useful to give some attention to it, to obtain a new insight into the possible limits of relevance of the theory of relativity, as well as into the relationship between relativity and quantum theory.

6 THE THEORY OF RELATIVITY

One of the first real breaks in classical notions of order and measure came with the theory of relativity. It is significant to point out here that the root of the theory of relativity was probably in a question that Einstein asked himself when he was fifteen years old: 'What would happen if one were to move at the speed of light and look in a mirror?' Evidently, one would see nothing because the light from one's face would never reach the mirror. This led Einstein to feel that light is somehow basically different from other forms of motion.

From our more modern vantage point, we can emphasize this difference yet more by considering the atomic structure of the matter out of which we are constituted. If we went faster than light, then, as a simple calculation shows, the electromagnetic fields that hold our atoms together would be left behind us (as the waves produced by an airplane are left behind it when it goes faster than sound). As a result, our atoms would disperse, and we would fall apart. So it would make no sense to suppose that we could go faster than light.

Now, a basic feature of the classical order and measure of Galileo and Newton is that one can in principle catch up with and overtake any form of motion, as long as the speed is finite. However, as has been indicated here, it leads to absurdities to suppose that we can catch up with and overtake light.

This perception that light should be considered to be different from other forms of motion is similar to Galileo's seeing that empty space and a viscous medium are different with regard to the expression of the laws of physics. In Einstein's case, one sees that the speed of light is not a possible speed for an object. Rather, it is like a horizon that cannot be reached. Even though we seem to move toward the horizon, we never get any closer. As we move toward a light ray, we never get closer to its speed. Its speed always remains the same, c, relative to us.

Relativity introduces new notions concerning the order and measure of time. These are no longer *absolute*, as was the case in Newtonian theory. Rather, they are now *relative* to the speed of a coordinate frame. This relativity of time is one of the radically new features of Einstein's theory.

A very significant change of language is involved in the expression of the new order and measure of time plied by relativistic theory. The speed of light is taken not as a possible speed of an *object*, but rather as the maximum speed of propagation of a *signal*. Heretofore, the notion of signal had played no role in the underlying general descriptive order of physics, but now it is playing a key role in this context.

The word 'signal' contains the word 'sign', which means 'to point to something' as well as 'to have significance'. A signal is indeed a kind of *communication*. So in a certain way, significance, meaning, and communication became relevant in the expression of the general descriptive order of physics (as did also information, which is, however, only a *part* of the content or meaning of a communication). The full implications of this have perhaps not yet been realized, i.e., of how certain very subtle notions of order going far beyond those of classical mechanics have tacitly been brought into the general descriptive framework of physics.

The new order and measure introduced in relativity theory implies new notions of structure, in which the idea of a rigid body can no longer play a key role. Indeed, it is not possible in relativity to obtain a consistent definition of an extended rigid body, because this would imply signals faster than light. In order to try to accommodate this new feature of relativity theory within the older notions of structure, physicists were driven to the notion of a particle that is an extensionless point, but, as is well known, this effort has not led to generally satisfactory results, because of the infinite fields implied by point particles. Actually, relativity implies that neither the point particles nor the quasi-rigid body

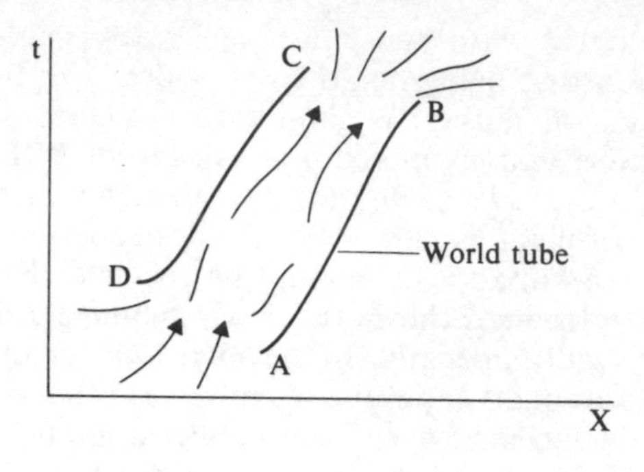

Figure 5.5

can be taken as primary concepts. Rather, these have to be expressed in terms of *events* and *processes*.

For example, any localizable structure may be described as a *world tube* (see Figure 5.5). Inside this tube *ABCD*, a complex process is going on, as indicated by the many lines within the world tube. It is not possible consistently to analyse movement within this tube in terms of 'finer particles' because these, too, would have to be described as tubes, and so on *ad infinitum*. Moreover, each tube is brought into existence from a broader background or context, as indicated by the lines preceding *AD*, while eventually it dissolves back into the background, as indicated by the lines following *BC*. Thus, the 'object' is an abstraction of a relatively invariant form. That is to say, it is more like a pattern of movement than like a solid separate thing that exists autonomously and permanently.[3]

However, thus far the problem of obtaining a *consistent* description of such a world tube has not been solved. Einstein did in fact very seriously try to obtain such a description in terms of a unified field theory. He took the total field of the whole universe as a primary description. This field is continuous and indivisible. Particles are then to be regarded as certain kinds of abstraction from the total field, corresponding to regions of very intense field (called singularities). As the distance from the singularity increases (see Figure 5.6), the field gets weaker, until it merges imperceptibly with the fields of other singularities. But nowhere is there a break or a division. Thus, the classical idea of the separability of the world into distinct but interacting parts is no

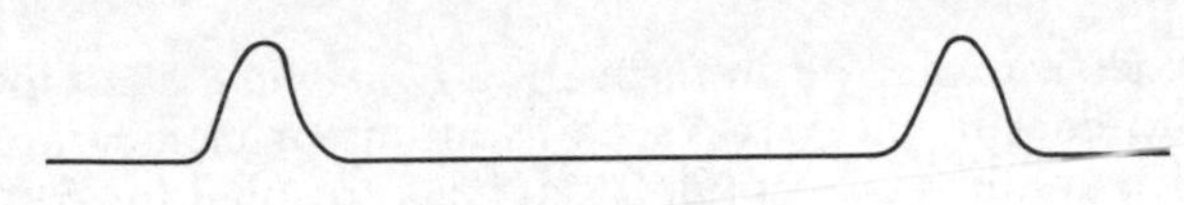

Figure 5.6

longer valid or relevant. Rather, we have to regard the universe as *an undivided and unbroken whole*. Division into particles, or into particles and fields, is only a crude abstraction and approximation. Thus, we come to an order that is radically different from that of Galileo and Newton – the order of *undivided wholeness*.

In formulating his description in terms of a unified field, Einstein developed *the general theory of relativity*. This involved a number of further new notions of order. Thus, Einstein considered arbitrary sets of *continuous curves* as allowable coordinates, so that he worked in terms of *curvilinear order and measure* rather than in terms of *rectilinear order and measure* (though of course such curves are locally still approximately rectilinear over short enough distances). Through the princples of equivalence of gravitation and acceleration and through the use of the Christoffel symbol $\Gamma^{a}{}_{bc}$ which mathematically describes the local rate of 'turning' of the curvilinear coodinates, Einstein was able to relate this curvilinear order and measure to the *gravitational field*. This relationship implied the need for *non-linear equations*, i.e., equations whose solution cannot simply be added together to yield new solutions. This non-linear feature of the equations was of crucial significance not only in that it in principle opened up the possibility of solutions with stable particle-like singularities of the type described above (which are impossible with linear equations), but also in that it had very important implications with regard to the question of *analysis* of the world into distinct but interacting components.

In discussing this question, it is useful first to note that the word 'analysis' has the Greek root 'lysis', which is also the root of the English 'loosen' and which means 'to break up or dissolve'. Thus, a chemist can break up a compound into its basic elementary constituents, and then he can put these constituents back together again, and thus *synthesize* the compound. The words 'analysis' and 'synthesis' have, however, come to refer not merely to actual physical or chemical operations with *things*, but also to similar operations carried out in *thought*. Thus, it may be said that classical physics is expressed in terms of a *conceptual analysis* of the world into constituent parts (such as atoms or elementary particles) which are then conceptually put back together to 'synthesize' a total system, by considering the interactions of these parts.

Such parts may be separate in space (as are the atoms), but they may also involve more abstract notions that do not imply separation in space. For example, in a wave field that satisfies a linear equation, it is possible to choose a set of 'normal modes' of motion of the entire field, each of which can be regarded as moving independently of the others. One can then *think* of the field analytically as if every possible form of wave motion were constituted out of a sum of such independent 'normal modes'. Even if the wave field satisfies a non-linear equation, one can in a certain approximation still analyse it in terms of a set of such 'normal modes', but these have now to be regarded as being mutually dependent because of a certain kind of interaction. However, this kind of 'analysis and synthesis' is of only limited validity because in general the solutions of non-linear equations have properties that cannot be expressed in terms of such an analysis. (In mathematical terms, it can be said, for example, that the analysis involves series that do not always converge.) Indeed, the non-linear equations of unified field theory are in general of this character. Thus, it is clear that not only is the notion of analysis in terms of spatially separate objects generally irrelevant in the context of such theories, but so also is the notion of analysis into more abstract constituents that are not regarded as separate in space.

It is important here to call attention to the difference between analysis and *description*. The word 'de-scribe' literally means to '*write* down', but when we write things down, this does not in general mean that the terms appearing in such a description can be actually 'loosened' or 'separated' into autonomously behaving components, and then put back together again in a synthesis. Rather, these terms are in general abstractions which have little or no meaning when considered as autonomous and separate from each other. Indeed, what is primarily relevant in a description is how the terms are *related* by ratio or reason. It is this ratio or reason, which calls attention to the whole, that is *meant* by a description.

Thus, even conceptually, a description does not in general constitute an analysis. Rather, a conceptual analysis provides a *special sort* of description, in which we can think about something as if it were broken into autonomously behaving parts, which are then thought about as put back together again in interaction. Such analytic forms of description were generally adequate for the physics of Galileo and Newton, but as has been indicated here, they have ceased to be so in the physics of Einstein.

Although Einsten made a very promising start along this new direction of thinking in physics, he was never able to arrive at a generally coherent and satisfactory theory, starting from the concept of a unified field. As pointed out earlier, physicists were therefore left with the problem of trying to adapt the older concept of analysis of the world into extensionless particles to the context of relativity, in which such an analysis of the world is not really relevant or consistent.

It will be helpful here to consider certain possible inadequacies in Einstein's approach to these questions, though of course only in a very preliminary way. In this connection, it is useful to recall that in 1905 Einstein wrote three very fundamental papers, one on relativity, one on the quantum of light (photoelectric effects) and one on Brownian motion. A detailed study of these papers shows that they are intimately related in a number of ways, and this suggests that in Einstein's early thinking he was at least tacitly regarding these three subjects as aspects of one broader unity. However, with the development of general relativity there came a very heavy emphasis on the *continuity of fields*. The other two subjects (Brownian motion and the quantum properties of light) which involved some kind of discontinuity that was not in harmony with the notion of a continuous field, tended to fall into the background, and eventually, to be more or less dropped from consideration, at least within the context of general relativity.

In discussing this question, it will be helpful first to consider Brownian motion, which is indeed very difficult to describe in a relativistically invariant way. Because Brownian motion implies infinite 'instantaneous velocities', it cannot be restricted to the speed of light. However, in compensation, Brownian motion cannot in general be the carrier of a signal, for a signal is some *ordered* modulation of a 'carrier'. This order is not separable from the *meaning* of the signal (i.e., to change the order is to change the meaning). Thus, one can properly speak of propagation of a signal only in a context in which the movement of the 'carrier' is so regular and continuous that the order is not mixed up. With Brownian motion, however, the order is of such a high degree (i.e., 'random' in the usual sense of the word) that the meaning of a signal would no longer be left unaltered in its propagation. Therefore, there is no reason why a Brownian curve of infinite order cannot be taken as part of a primary description of movement, as long as its *average* speed is not greater than that of light. In this way, it is possible for relativity theory to emerge as relevant to the *average speed* of a Brownian curve (which would also be appropriate for discussing the propagation of a signal), while it would have no relevance in a broader context in which the primary law would relate to Brownian curves of indefinitely high degree, rather than to a continuous curve of low degree. To develop such a theory would evidently imply a new order and measure in physics (going beyond both Newtonian and Einsteinian ideas), and it would lead to correspondingly new structures.

Consideration of such notions may perhaps point to something new and relevant. However, before this sort of inquiry is pursued further, it is better to go into the quantum theory, which is in many ways even more significant in this context than is Brownian motion.

7 QUANTUM THEORY

The quantum theory implies a much more radical change in notions of order and measure than even relativity did. To understand this change, one has to consider four new features of primary significance introduced by this theory.

7.1 Indivisibility of the Quantum of Action

This indivisibility implies that transitions between stationary states are in some sense discrete. Thus, it has no meaning to say that a system passes through a continuous series of intermediate states, similar to initial and final states. This is, of course, quite different from classical physics, which implies such a continuous series of intermediate states in every transition.

7.2 Wave–Particle Duality of the Properties of Matter

Under different experimental conditions, matter behaves more like a wave or more like a particle, but always, in certain ways, like both together.

7.3 Properties of Matter as Statistically Revealed Potentialities

Every physical situation is now characterized by a wave function (or more abstractly by a vector in Hilbert space). This wave function is not directly related to the *actual* properties of an individual object, event, or process. Rather, it has to be thought of as a description of the *potentialities* within the physical situation.[4] Different and generally mutually incompatible potentialities (e.g., for wavelike or particle-like behaviour) are actualized in different experimental arrangements (so that the wave-particle duality can be understood as one of the main forms for the expression of such incompatible potentialities). In general, the wave function gives only a *probability measure* for the actualization of different potentialities in a statistical ensemble of similar observations carried out under specified conditions, and cannot predict what will happen in detail in each individual observation.

This notion of statistical determination of mutually incompatible potentialities is evidently very different from what is done in classical physics, which has no place in it to give the notion of potentiality such a fundamental role. In classical physics, one thinks that only the *actual state* of a system can be relevant in a given physical situation, and that probability comes in either because we are ignorant of the actual state or because we are averaging over an ensemble of actual states that are distributed over a range of conditions. In quantum theory it has no meaning to discuss the actual state of a system apart from the whole set of experimental conditions which are essential to *actualize* this state.

7.4 Non–causal Correlations (the Paradox of Einstein, Podolsky and Rosen)

It is an inference from the quantum theory that events that are separated in space and that are without possibility of connection through interaction are correlated, in a way that can be shown to be incapable of a detailed causal explanation, through the propagation of effects at speeds not greater than that of light.[5] Thus, the quantum theory is not compatible with Einstein's basic approach to relativity, in which it is essential that such correlations be explainable by signals propagated at speeds not faster than that of light.

All of these evidently imply a breakdown of the general order of description that had prevailed before the advent of quantum theory. The limits of this 'pre-quantum' order are indeed brought out very clearly in terms of the uncertainty relations, which are commonly illustrated in terms of Heisenberg's famous microscope experiment.

This experiment will now be discussed here, in a form somewhat different from that used by Heisenberg, in order to bring out certain new points. Our first step is to go into what it means to make a *classical* measurement of position and momentum. In doing this, we consider the use of an *electron* microscope, rather than a *light* microscope.

As shown in Figure 5.7, there is in the target an 'observed particle' at *O*, assumed to have initially a known momentum (e.g., it may be at rest, with zero momentum). Electrons of known energy are incident on the target, and one of these is deflected by the particle at *O*. It goes through the electron lens, following an orbit that leads it to the focus at *P*. From here, the electron leaves a track *T* in a certain direction, as it penetrates the photographic emulsion.

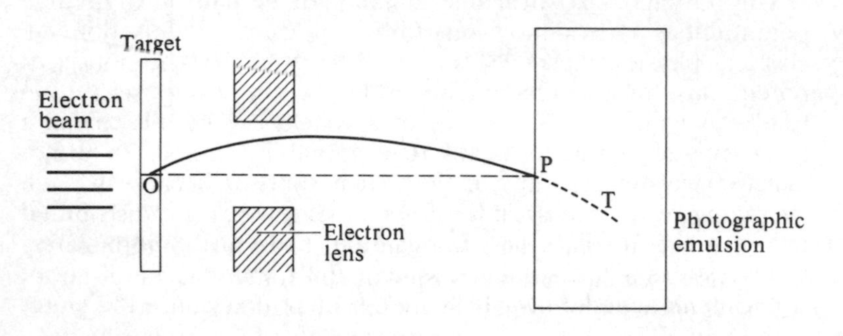

Figure 5.7

Now, the *directly observable results* of this experiment are the position *P* and the direction of the track *T*, but of course these are in themselves of no interest. It is only by knowing overall the experimental conditions (i.e., the structure of the microscope, the target, the energy of the incident beam of electrons, etc.) that the experimental results become significant in the context of a physical inquiry. With the aid of an adequate description of these conditions, one can use the experimental results to make inferences about the position of the 'observed particle' at *O*, and about the momentum transferred to it in the process of deflecting the incident electron. Thus, although the operation of the instrument does influence the observed particle, this influence can be taken into account, so that we can infer, and thus 'know', both the position and momentum of this particle at the time of deflection of the incident electron.

All this is quite straightforward in the context of classical physics. Heisenberg's novel step was to consider the implications of the 'quantum' character of the electron that provides the 'link' between the *experimental results* and *what is to be inferred from these results*. This electron can no longer be described as being just a classical particle. Rather, it has also to be described in terms of a 'wave', as shown in Figure 5.8. Electron waves are said to be incident on the target, and diffracted by the atom at O.

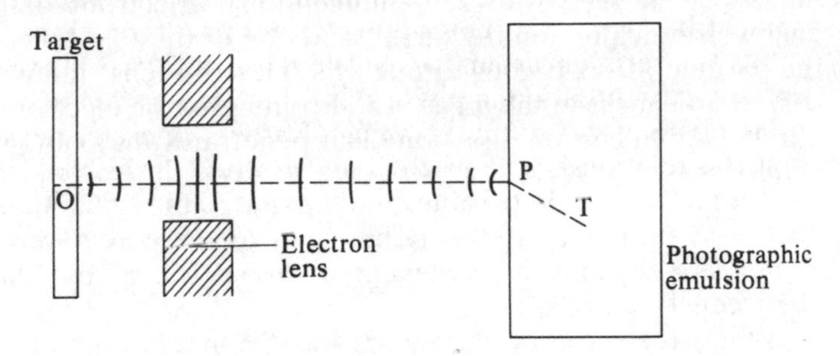

Figure 5.8

They then pass through the lens, where they are further diffracted and brought to focus in the emulsion at P. From here, there starts a track T (just as happened in the classical description).

Evidently, Heisenberg has brought in the four primarily significant features of the quantum theory referred to at the beginning of this section. Thus (as happens in the interference experiment also), he describes the link electron *both* as a wave (while it is passing from object O through the lens to the image at P) *and* as a particle (when it arrives at the point P and then leaves a track T). The transfer of momentum to the 'observed atom' at O has to be treated as discrete and indivisible. Between O and P the most detailed possible description of the link electron is in terms of a wave function that determines only a statistical distribution of potentialities whose actualization depends on the experimental conditions (e.g., the presence of sensitive atoms in the emulsion, which can reveal the electron). Finally, the actual results (the spot P, the track T, and the properties of the atom O) are correlated in the non-causal way mentioned earlier in this chapter.

By using all these primary features of the quantum theory in discussing the 'link' electron, Heisenberg was able to show that there is a limit to the precision of inferences that can be drawn about the observed object, given by the uncertainty relations ($\Delta x \times \Delta p \geqslant h$). At first, Heisenberg explained the uncertainty as the result of the 'uncertain' character of the precise orbit of the 'electron link' between O and P, which also implied an uncertain 'disturbance' of the atom O when this electron was scattered. However, Bohr[6] gave a relatively thorough and consistent discussion of the whole situation, which made it clear that the four primary aspects of the quantum theory as described above are not compatible with any description in terms of precisely defined orbits that are 'uncertain' to us. We have thus to do here with an entirely new situation in physics, in which the notion of a detailed orbit no longer has any meaning. Rather, one can perhaps say that the relationship between O and P through the 'link' electron is similar to an indivisible and unanalysable 'quantum jump' between stationary states, rather than to the continuous though not precisely known movement of a particle across the space between O and P.

What, then, can be the significance of the description that has been given of Heisenberg's experiment? Evidently, it is only in a context in which classical physics is applicable that this experiment can coherently be discussed in this way. Such a discussion can therefore at most serve to indicate the *limits of relevance* of classical modes of description; it cannot actually provide a description that is coherent in a 'quantum' context.

Even when regarded in this way, however, the usual discussion of the experiment overlooks certain key points, which have deep and far-reaching significance. To see what these are, we note that from a particular set of experimental conditions as determined by the structure of the microscope, etc., one could in some rough sense say that the limits of applicability of the classical description are indicated by a certain cell in the phase space of this object, which we describe by A in Figure 5.9. If, however, there had been a different set of experimental conditions (e.g., a microscope of another aperture, electrons of different energy, etc.), then these limits would have had to be indicated by another cell in phase space, indicated by B. Heisenberg emphasized that both cells must

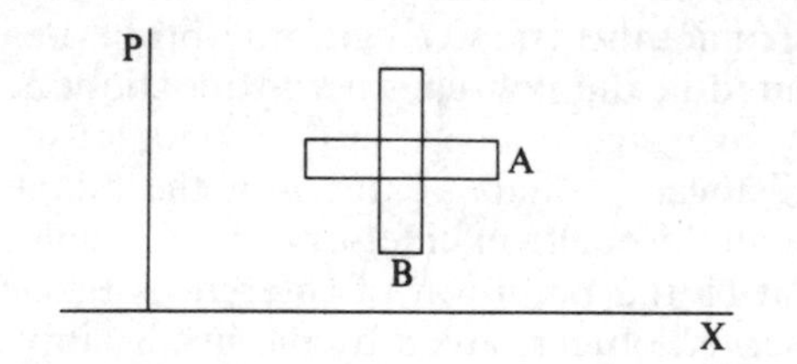

Figure 5.9

have the same area, h, but in doing this he left out of account the significance of the fact that their 'shapes' are different.

Of course, in the context of classical physics (in which quantities of the order of Planck's constant, h, can be neglected), all cells can be replaced by dimensionless points, so that their 'shapes' have no significance at all. Therefore, the experimental results can be said to do nothing more than permit inferences to be drawn about an observed object, inferences in which the 'shapes' of the cells, and therefore the details of the experimental conditions, play only the role of intermediary links in the chain of reasoning, which drop out of the ultimate result that is inferred. This means that the observed object can consistently be said to exist separately and independently of the observing instrument, in the sense that it can be regarded as 'having' certain properties whether it interacts with anything else (such as an observing instrument) or not.

However, in the 'quantum' context the situation is very different. Here, the 'shapes' of the cells remain relevant, as essential parts of the description of the observed particle. This latter therefore cannot properly be described except in conjunction with a description of the experimental conditions; and if one goes in more detail into a mathematical treatment according to the laws of the quantum theory, the 'wave function' of the 'observed object' cannot be specified apart from a specification of the wave function of the 'link electron', which in turn requires a description of the overall experimental conditions (so that the relationship between the object and the observed result is actually an example of the correlations of the type indicated by Einstein, Podolsky and Rosen, which cannot be explained in terms of the propagation of signals as chains of causal influence). This means that the description of the experimental conditions does not drop out as a mere intermediary link of inference, but remains inseparable from the description of what is called the observed object. The 'quantum' context thus calls for a new kind of description that does not imply the separability of the 'observed object' and 'observing instrument'. Instead, the form of the experimental conditions and the meaning of the experimental results have now to be one whole, in which analysis into autonomously existent elements is not relevant.

What is meant here by wholeness could be indicated metaphorically by calling attention to a pattern (e.g., in a carpet). In so far as what is relevant *is* the pattern, it has no meaning to say that different parts of such a pattern (e.g., various flowers and trees that are to be seen in the carpet) are separate objects in interaction. Similarly, in the quantum context, one can regard terms like 'observed object', 'observing instrument', 'link electron', 'experimental results', etc., as aspects of a single overall 'pattern' that are in effect abstracted or 'pointed out' by our mode of description. Thus, to speak of the interaction of 'observing instrument' and 'observed object' has no meaning.

A centrally relevant change in descriptive order required in the quantum theory is thus the dropping of the notion of analysis of the world into relatively autonomous parts, separately existent but in interaction. Rather, the primary emphasis is now on *undivided wholeness*, in which the observing instrument is not separable from what is observed.

Though quantum theory is very different from relativity, yet in some deep sense they have in common this implication of undivided wholeness. Thus, in relativity, a consistent description of the instruments would have to be in terms of a structure of singularities in the field (corresponding to what are now generally called 'the constituent atoms' of the instrument). These would merge with the fields of the singularities constituting the 'observed particle' (and eventually with those constituting 'the atoms out of which the human observer is constituted'). This is a different sort of wholeness from that implied by the quantum theory, but it is similar in that there can be no ultimate division between the observing instrument and the observed object.

Nevertheless, in spite of this deep similarity, it has not proved possible to unite relativity and quantum theory in a coherent way. One of the main reasons is that there is no consistent means of introducing extended structure in relativity, so that particles have to be treated as extensionless points. This has led to infinite results in quantum field-theoretical calculations. By means of various formal algorithms (e.g., renormalization, *S* matrices, etc.) certain finite and essentially correct results have been abstracted from the theory. However, at bottom, the theory remains generally unsatisfactory, not only because it contains what at least appear to be some serious contradictions, but also because it certainly has a number of arbitrary features which are capable of indefinite adaptation to the facts, somewhat reminiscent of the way in which the Ptolemaic epicycles could be made to accommodate almost any observational data that might arise in the application of such a descriptive framework (e.g., in renormalization, the vacuum-state wave function has an infinite number of arbitrary features).

It would not, however, be very helpful here to make a detailed analysis of these problems. Rather, it will be more useful to call attention to a few general difficulties, the consideration of which will perhaps show that these details are not very relevant in the context of the present discussion.

First, quantum field theory begins by defining a field $\Psi(\mathbf{x}, t)$. This field is a quantum operator, but $\mathbf{x}$ and t describe a continuous order in space and time. To bring the point out in more detail, we can write the matrix element $\Psi_{ij}(\mathbf{x}, t)$. However, as soon as we impose relativistic invariance, we deduce 'infinite fluctuations', i.e., $\Psi_{ij}(\mathbf{x}, t)$ is in general infinite and discontinuous because of 'zero-point' quantum fluctuations. This contradicts the original assumption of continuity of all functions required in any relativistic theory.

This emphasis on continuous orders is (as has been pointed out in the previous section) a serious weakness of the theory of relativity. If we deal with discontinuous order, however (e.g., as in Brownian motion), then the notion of signal ceases to be relevant (and with it, the notion of limitation to the speed of light); and without the notion of signal in a basic role, we are once again free to consider extended structures in a primary role in our descriptions.

Of course, the limitation to the speed of light will hold on the average and in the long run. Thus, relativistic notions will be relevant in suitable limiting cases. But the theory of relativity need not just be imposed on quantum theory. It is this imposition of the underlying *descriptive order* of one theory on another that led to arbitrary features and possible contradictions.

To see how this comes about we note that if the relativistic notion of giving a fundamental role to the possibility of *signalling* from one region *point* to another is to have any meaning, the *source* of a signal must be clearly separated from the region in which it is *received*, not only spatially but also in the sense that the two must be essentially autonomous in their behaviour.

Thus, as shown in Figure 5.10, if a signal is emitted from the world tube of a source of A, then it has to be propagated continuously without change of order to B, the world tube of the receiver. However, at a quantum level of description, the time order of events in the world tube at A and B may, according to the uncertainty principle, cease to be definable in the usual way. This alone would make the notion of a signal meaningless. In addition, the notion of a clear and distinct spatial separation of A and B, as well as that of possible autonomy in their behaviour, will cease to be relevant, because the 'contact' between A and B has now to be regarded as similar to an indivisible quantum jump of an atom between stationary states. Moreover, the further development of this notion along the lines of the experiment of Einstein, Podolsky and Rosen leads to the inference that the connection between *A* and *B* cannot in general be described in terms of the propagation of causal influences (which type of propagation is evidently necessary to provide for an underlying 'carrier' of the signal).

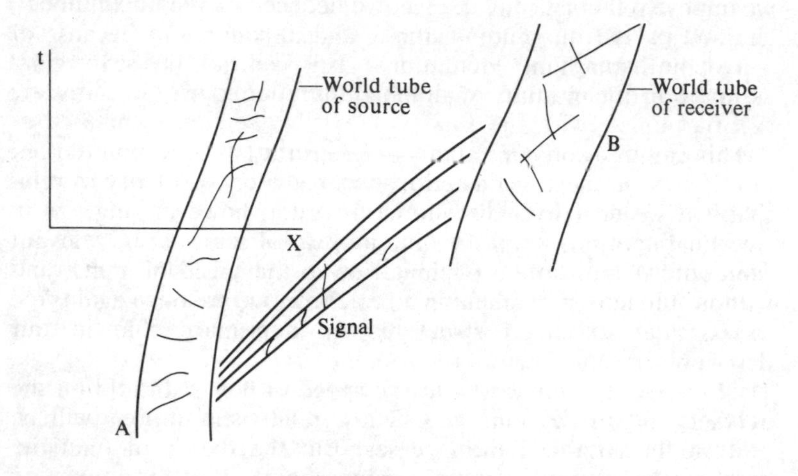

Figure 5.10

It seems clear, then, that the relativistic notion of a signal simply does not fit coherently into the 'quantum' context. This is basically because such a signal implies the possibility of a *certain kind of analysis* which is not compatible with the sort of undivided wholeness that is implied by the quantum theory. In this connection, it may indeed be said that although Einstein's unified field theory denies the possibility of ultimate analysis of the world into autonomous component elements, nevertheless, the notion that the possibility of a signal plays such a basic role implies a different and more abstract sort of analysis based on a kind of independent and autonomous 'information content' which is different in different regions. This abstract kind of analysis may not only be inconsistent with quantum theory but, very probably, also with the undivided wholeness implied in the other aspects of the theory of relativity.

What suggests itself, then, is that we seriously consider the possibility of dropping the idea of the basic role of the notion of signal, but go on with the other aspects of relativity theory (especially the principle that laws are invariant relationships, and that through non-linearity of the equations, or in some other way, analysis into autonomous components will cease to be relevant). Thus, by letting go of this kind of attachment to a certain kind of analysis that does not harmonize with the 'quantum' context, we open the way for a new theory that comprehends what is still valid in relativity theory, but does not deny the indivisible wholeness implied by the quantum theory.

On the other hand, quantum theory also contains an implicit attachment to a certain very abstract kind of analysis, which does not harmonize with the sort of indivisible wholeness implied by the theory of relativity. To see what this is, we note that discussions such as those centring around the Heisenberg microscope emphasize the indivisible wholeness of the observing instrument and the observed object only in context of the *actual* results of an experiment. However, in the mathematical theory, the wave function is still generally taken to be a description of overall *statistical potentialities* that are regarded as existing separately and autonomously. In other words, the *actual and individual object* of classical physics is replaced by a more abstract kind of *potential and statistical object*. This latter is said to correspond to the 'quantum state of the system', which in turn corresponds to 'the wave function of the system' (or more generally to a vector in Hilbert space). Such usage of language (e.g., to bring in words such as 'state of a system') implies that we are thinking of something that has a separate and autonomous kind of existence.

The consistency of this way of using language depends to a large extent on the mathematical assumption that the wave equation (i.e., the law governing the changes with time of the wave function, or the Hilbert space vector) is linear. (Non-linear equations for field operations have been proposed but, even here, this is only a limited kind of non-linearity, in the sense that the basic equation for 'the state vector in Hilbert space' is always taken to be linear.) Such linearity of equations then allows us to regard 'state vectors' as having a kind of autonomous existence (similar in certain ways to that which is attributed in classical field theories to normal modes, but different in that they are more abstract).

This complete autonomy of the 'quantum state' of a system is supposed to hold only when it is not being observed. In an observation, it is assumed that we have to do with two initially autonomous systems that have come into interaction.[7] One of these is described by the 'state vector of the observed object' and the other by the 'state vector of the observing apparatus'.

In the consideration of this interaction, certain new features are introduced, which correspond to allowing for the possibility of actualizing the observed system's potentialities at the expense of others that cannot be actualized at the same time. (Mathematically, one can say that 'the wave packet is reduced' or that 'a projection operation takes place'.)

There is a great deal of controversy and discussion as to precisely how this stage is to be treated, because the basic notions involved do not seem to be very clear. However, it is not our aim here to criticize these efforts in detail. Rather, we wish merely to point out that this whole line of approach re-establishes at the abstract level of statistical potentialities the same kind of analysis into separate and autonomous components in interaction that is denied at the more concrete level of individual objects. It is just this kind of abstract analysis that does not cohere with the underlying basic descriptive order of relativity theory, for, as has been seen, relativity theory is not compatible with such an analysis of the world into separate components. Rather, it ultimately implies that such 'objects' have to be understood as merging with each other (as field singularities do) to make one indivisible whole. Similarly, one may consider the notion that through a thoroughgoing non-linearity, or in some other way, quantum theory may be allowed to change, so that the resulting new theory will also imply undivided wholeness, not merely at the level of actual individual phenomena, but also at the level of potentialities treated in terms of statistical aggregates. In this way, those aspects of quantum theory that are still valid will be able to harmonize with those aspects of relativity that are still valid.

To give up both the basic role of signal and that of quantum state is, however, no small thing. To find a new theory that goes on without these will evidently require radically new notions of order, measure and structure.

One may suggest here that we are in a position which is in certain ways similar to where Galileo stood when he began his inquiries. A great deal of work has been done showing the inadequacy of old ideas, which merely permit a range of new facts to be *fitted mathematically* (comparable to what was done by Copernicus, Kepler and others), but we have not yet freed ourselves thoroughly from the old order of thinking, using language, and observing. We have thus yet to perceive a new *order*. As with Galileo, this must involve seeing new differences so that much of what has been thought to be basic in the old ideas will be perceived be perceived to be more or less correct, but not of primary relevance (as happened, for example, with some of the key ideas of Aristotle). When we see the new basic differences, then (as happened with Newton) we will be able to perceive a new universal ratio or reason relating and unifying all the differences. This may ultimately carry us as far beyond quantum theory and relativity as Newton's ideas went beyond those of Copernicus.

Of course, this cannot be done overnight. We have to work patiently, slowly, and carefully, to understand the present general situation in physics in a new way. Some preliminary steps of this kind will be discussed in chapter 6.

6

Quantum theory as an indication of a new order in physics
Part B: Implicate and explicate order in physical law

1 INTRODUCTION

Chapter 5 called attention to the emergence of new orders throughout the history of physics. A general feature of the development of this subject has been a tendency to regard certain basic notions of order as permanent and unchangeable. The task of physics was then taken to be to *accommodate* new observations by means of *adaptations* within these basic notions of order, so as to fit the new facts. This kind of adaptation began with the Ptolemaic epicycles, which continued from ancient times until the advent of the work of Copernicus, Kepler, Galileo, and Newton. As soon as the basic notions of order in classical physics had been fairly clearly expressed, it was supposed that further work in physics would consist of adaptation within this order to accommodate new facts. This continued until the appearance of relativity and the quantum theory. It can accurately be said that since then the main line of work in physics has been adaptation within the general orders underlying these theories, to accommodate the facts to which these in turn have led.

It may thus be inferred that accommodation within already existing frameworks of order has generally been considered to be the main activity to be emphasized in physics, while the perception of new orders has been thought of as something that happens only occasionally, perhaps in revolutionary periods, during which what is regarded as the normal process of accommodation has broken down.[1]

It is pertinent to this subject to consider Piaget's[2] description of all intelligent perception in terms of two complementary movements, *accommodation* and *assimilation*. From the roots 'mod', meaning 'measure', and 'com', meaning 'together', one sees that to accommodate means 'to establish a common measure' (see chapter 5 for a discussion of the broader sense of the notion of measure that are relevant in this context). Examples of accommodation are fitting, cutting to a pattern, adapting, imitating, conforming to rules, etc. On the other hand, 'to assimilate' is 'to digest' or to make into a comprehensive and inseparable whole (which includes oneself). Thus, to assimilate means 'to understand'.

It is clear that in intelligent perception, primary emphasis has in general to be given to assimilation, while accommodation tends to play a relatively secondary role in the sense that its main significance is as an aid to assimilation.

Of course, we are able in certain sorts of contexts just to accommodate something that we observe within known orders of thought, and in this very act it will be adequately assimilated. However, it is necessary in more general contexts to give serious attention to the possibility that the old orders of thought may cease to be relevant, so that they can no longer coherently be adapted to fit the new fact. As has been brought out in some detail in chapter 5 one may then have to see the irrelevance of old differences, and the relevance of new differences, and thus one may open the way to the perception of new orders, new measures and new structures.

Clearly, such perception can appropriately take place at almost any time, and does not have to be restricted to unusual and revolutionary periods in which one finds that the older orders can no longer be conveniently adapted to the facts. Rather, one may be continually ready to drop old notions of order at various contexts, which may be broad or narrow, and to perceive new notions that may be relevant in such contexts. Thus, understanding the fact by assimilating it into new orders can become what could perhaps be called the normal way of doing scientific research.

To work in this way is evidently to give primary emphasis to something similar to *artistic perception*. Such perception begins by observing the whole fact in its full individuality, and then by degree articulates the order that is proper to the assimilation of this fact. It does not begin with abstract preconceptions as to what the order has to be, which are then adapted to the order that is observed.

What, then, is the proper role of accommodation of facts within known theoretical orders, measures and structures? Here, it is important to note that facts are not to be considered as if they were independently existent objects, that we might find or pick up in the laboratory. Rather, as the Latin root of the word 'facere' indicates, the fact is 'what has been made' (e.g., as in 'manufacture'). Thus, in a certain sense, we 'make' the fact. That is to say, beginning with immediate perception of an actual situation, we develop the fact by giving it further order, form and structure with the aid of our theoretical concepts. For example, by using the notions of order prevailing in ancient times, men were led to 'make' the fact about planetary motions by describing and measuring in terms of epicycles. In classical physics, the fact was 'made' in terms of the order of planetary orbits, measured through positions and times. In general relativity, the fact was 'made' in terms of the order of Riemannian geometry, and of the measure implied by concepts such as 'curvature of space'. In the quantum theory, the fact was made in terms of the order of energy levels, quantum numbers, symmetry groups, etc., along with appropriate measures (e.g. scattering cross-sections, charges, and masses of particles, etc.).

It is clear, then, that changes of order and measures in the theory ultimately lead to new ways of doing experiments and to new kinds of instruments, which in turn lead to the 'making' of correspondingly ordered and measured facts of new kinds. In this development, the experimental fact serves in the first instance as a test for theoretical notions. Thus, as has been pointed out in chapter 5, the general form of theoretical explanation is that of a generalized kind of ratio of reason. 'As A is to B in our structure of thinking, so it is in fact.' This ratio or reason constitutes a kind of 'common measure' or 'accommodation' between theory and fact.

As long as such a common measure prevails, then of course the theory used need not be changed. If the common measure is found not to be realized, then the first step is to see whether it can be re-established by means of adjustments within the theory without a change in its underlying order. If, after reasonable efforts, a proper accommodation of this kind is not achieved, then what is needed is a fresh perception of *the whole fact*. This now includes not only the results of experiments but also the *failure of certain lines of theory to fit the experimental results in a 'common measure'*. Then, as has been indicated earlier, one has to be very sensitively aware of all the relevant differences which underly the main orders in the old theory, to see whether there is room for a change of overall order. It is being emphasized here that this kind of perception should properly be interwoven continually with the activities aimed at accommodation, and should not have to be delayed for so long that the whole situation becomes confused and chaotic, apparently requiring the revolutionary destruction of the old order to clear it up.

As relativity and quantum theory have shown that it has no meaning to divide the observing apparatus from what is observed, so thc considerations discussed here indicate that it has no meaning to separate the observed fact (along with the instruments used to observe it) from the theoretical notions of order that help to give 'shape' to this fact. As we go on to develop new notions of order going beyond those of relativity and quantum theory, it will thus not be appropriate to try immediately to apply these notions to current problems that have arisen in the consideration of the present set of experimental facts. Rather, what is called for in this context is very broadly to assimilate the whole of the fact in physics into the new theoretical notions of order. After this fact has geneally been 'digested', we can begin to glimpse new ways in which such notions of order can be tested and perhaps extended in various directions. As pointed out at the end of chapter 5, we have to proceed slowly and patiently here or else we may become confused by 'undigested' facts.

Fact and theory are thus seen to be different aspects of one whole in which analysis into separate but interacting parts is not relevant. That is to say, not only is undivided wholeness implied in the *content* of physics (notably relativity and quantum theory) but also in the *manner of working* in physics. This means that we do not try *always* to force the theory to fit the kinds of facts that may be appropriate in currently accepted general orders of description, but that we are also ready when necessary to consider changes in what is meant by fact, which may be required for assimilation of such fact into new theoretical notions of order.

2 UNDIVIDED WHOLENESS – THE LENS AND THE HOLOGRAM

The undivided wholeness of modes of observation, instrumentation and theoretical understanding indicated above implies the need to consider a *new order of fact*, i.e., the fact about the way in which modes of theoretical understanding and of observation and instrumentation are related to each other. Until now, we have more or less just taken such a relationship for granted, without giving serious attention to the manner in which it arises, very probably because of the belief that the study of the subject belongs to 'the history of science' rather than to 'science proper'. However, it is now being suggested that the consideration of this relationship is essential for an adequate understanding of science itself, because the content of the observed fact cannot coherently be regarded as separate from modes of observation and instrumentation and modes of theoretical understanding.

An example of the very close relationship between instrumentation and theory can be seen by considering the *lens*, which was indeed one of the key features behind the development of modern scientific thought. The essential feature of a lens is, as indicated in Figure 6.1, that it forms an *image* in which a given point *P* in the object corresponds (in a high degree of approximation) to a point *Q* in the image. By thus bringing the correspondence of specified features of object and image into such sharp relief, the lens greatly strengthened man's awareness of the various parts of the object and of the relationship between these parts. In this way, it furthered the tendency to think in terms of analysis and synthesis. Moreover, it made possible an enormous extension of the classical order of analysis and synthesis to objects that were too far away, too big, too small, or too rapidly moving to be thus ordered by means of unaided vision. As a result, scientists were encouraged to extrapolate their ideas and to think that such an approach would be relevant and valid no matter how far they went, in all possible conditions, contexts, and degrees of approximation.

However, as has been seen in chapter 5 relativity and quantum theory imply undivided wholeness, in which analysis into distinct

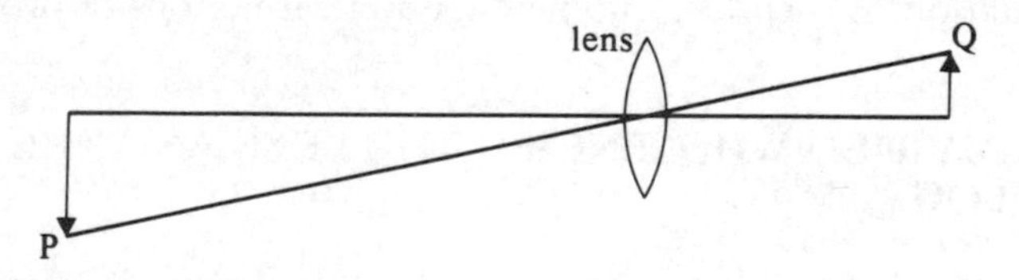

Figure 6.1

and well-defined parts is no longer relevant. Is there an instrument that can help give a certain immediate perceptual insight into what can be meant by individed wholeness, as the lens did for what can be meant by analysis of a system into parts? It is suggested here that one can obtain such insight by considering *hologram*. (The name is derived from the Greek words 'holo', meaning 'whole', and 'gram' meaning 'to write'. Thus, the hologram is an instrument that, as it were, 'writes the whole'.)

As shown in Figure 6.2 coherent light from a laser is passed through a half-silvered mirror. Part of the beam goes on directly to a photographic plate, while another part is reflected so that it illuminates a certain whole structure. The light reflected from this whole structure also reaches the plate, where it interferes with that arriving there by a direct path. The resulting interference pattern which is recorded on the plate is not only very complex but also usually so fine that it is not even visible to the naked eye. Yet, it is somehow relevant to the whole illuminated structure, though only in a highly implicit way.

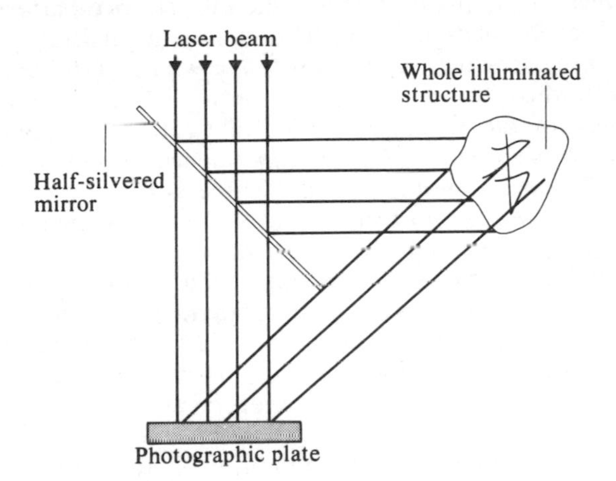

Figure 6.2

This relevance of the interference pattern to the whole illuminated structure is revealed when the photographic plate is illuminated with laser light. As shown in Figure 6.3, a wavefront is then created which is very similar in form to that coming off the original illuminated structure. By placing the eye in this way, one in effect sees the whole of the original structure, in three dimen-

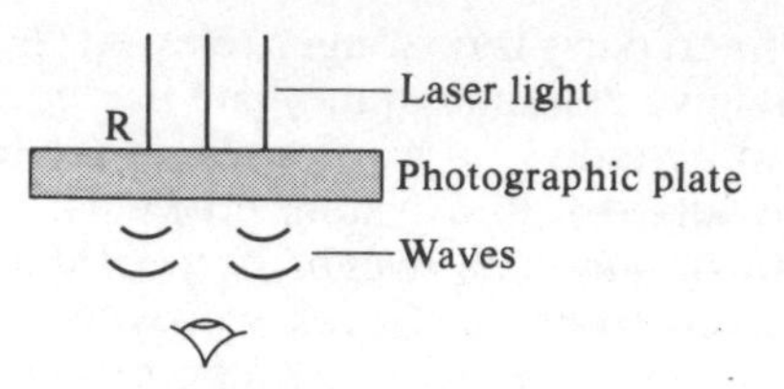

Figure 6.3

sions, and from a range of possible points of view (as if one were looking at it through a window). If we then illuminate only a small region R of the plate, we *still see the whole structure*, but in somewhat less sharply defined detail and from a decreased range of possible points of view (as if we were looking through a smaller window).

It is clear, then, that there is no one-to-one correspondence between parts of an 'illuminated object' and parts of an 'image of this object on the plate'. Rather, the interference pattern in each region R of the plate is relevant to the whole structure, and each region of the structure is relevant to the whole of the interference pattern on the plate.

Because of the wave properties of light, even a lens cannot produce an exact one-to-one correspondence. A lens can therefore be regarded as a limiting case of a hologram.

We can, however, go further and say that in their overall ways of indicating the meaning of observations, typical experiments as currently done in physics (especially in the 'quantum' context) are more like the general case of a hologram than like the special case of a lens. For example, consider a scattering experiment. As shown in Figure 6.4 what can be observed in the detector is generally relevant to the whole target, or at least to an area large enough to contain a great many atoms.

Moreover, although one might in principle try to make an image of a particular atom, the quantum theory implies that to do this would have little or no significance. Indeed, as the discussion of the Heisenberg microscope experiment given in chapter 5 shows, the formation of an image is just what is *not* relevant in a 'quantum' context; at most a discussion of image formation serves to indicate the limits of applicability of classical modes of description.

So we may say that in current research in physics, an instrument tends to be relevant to a whole structure, in a way rather similar

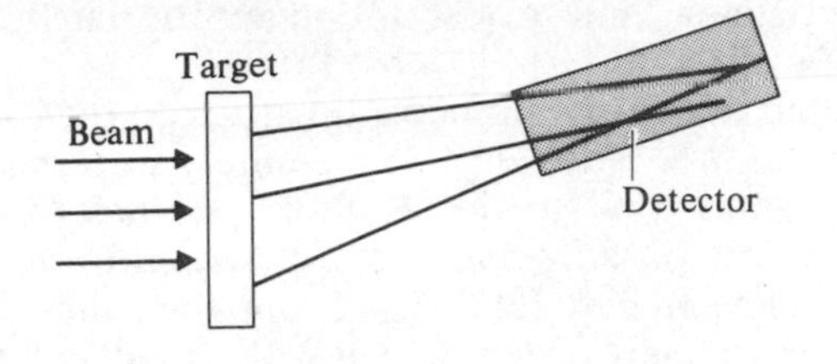

Figure 6.4

to what happens with a hologram. To be sure, there are certain differences. For example, in current experiments with electron beams or with X-rays, these latter are seldom coherent over appreciable distances. If, however, it should ever prove to be possible to develop something like an electron laser or an X-ray laser, then experiments will directly reveal 'atomic' and 'nuclear' structures without the need for complex chains of inference of the sort now generally required, as the hologram does for ordinary large-scale structures.

3 IMPLICATE AND EXPLICATE ORDER

What is being suggested here is that the consideration of the difference betwen lens and hologram can play a significant part in the perception of a new order that is relevant for physical law. As Galileo noted the distinction between a viscous medium and a vacuum and saw that physical law should refer primarily to the order of motion of an object in a vacuum, so we might now note the distinction between a lens and a hologram and consider the possibility that physical law should refer primarily to an order of undivided wholeness of the content of a description similar to that indicated by the hologram rather than to an order of analysis of such content into separate parts indicated by a lens.

However, when Aristotle's ideas on movement were dropped, Galileo and those who followed him had to consider the question of how the new order of motion was to be described in adequate details. The answer came in the form of Cartesian coordinates extended to the language of the calculus (differential equations, etc). But this kind of description is of course appropriate only in a context in which analysis into distinct and autonomous parts is relevant, and will therefore in turn have to be dropped. What, then, will be the new kind of description appropriate to the present context?

As happened with Cartesian coordinates and the calculus, such a question cannot be answered immediately in terms of definite prescriptions as to what to do. Rather, one has to observe the new situation very broadly and tentatively and to 'feel out' what may be the relevant new features. From this, there will arise a discernment of the new order, which will articulate and unfold in a natural way (and not as a result of efforts to make it fit well-defined and preconceived notions as to what this order should be able to achieve).

We can begin such an inquiry by noting that in some subtle sense, which does not appear in ordinary vision, the interference pattern in the whole plate can distinguish different orders and measures in the whole illuminated structure. For example, the illuminated structure may contain all sorts of shapes and sizes of

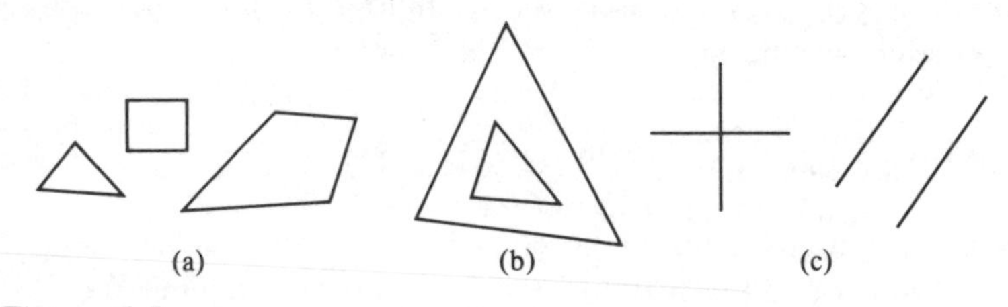

Figure 6.5

geometric forms (indicated in Figure 6.5a), as well as topological relationships, such as inside and outside (indicted in Figure 6.5b), and intersection and separation (indicated in Figure 6.5c). All of these lead to different interference patterns and it is this difference that is somehow to be described in detail.

The differences indicated above are, however, not only in the plate. Indeed, the latter is of secondary significance, in the sense that its main function is to make a relatively permanent 'written record' of the interference pattern of the light that is present in each region of space. More generally, however, in each such region, the movement of the light implicitly contains a vast range of distinctions of order and measure, appropriate to a whole illuminated structure. Indeed, in principle, this structure extends over the whole universe and over the whole past, with implications for the whole future. Consider, for example, how on looking at the night sky, we are able to discern structures covering immense stretches of space and time, which are in some sense contained in the movements of light in the tiny space encompassed by the eye (and also how instruments, such as optical and radio telescopes, can discern more and more of this totality, contained in each region of space).

There is the germ of a new notion of order here. This order is not to be understood solely in terms of a regular arrangement of *objects* (e.g., in rows) or as a regular arrangement of *events* (e.g. in a series). Rather, a *total order* is contained, in some *implicit* sense, in each region of space and time.

Now, the word 'implicit' is based on the verb 'to implicate'. This means 'to fold inward' (as multiplication means 'folding many times'). So we may be led to explore the notion that in some sense each region contains a total structure 'enfolded' within it.

It will be useful in such an exploration to consider some further examples of enfolded or *implicate* order. Thus, in a television broadcast, the visual image is translated into a time order, which is 'carried' by the radio wave. Points that are near each other in the visual image are not necessarily 'near' in the order of the radio signal. Thus, the radio wave carries the visual image in an implicate order. The function of the receiver is then to *explicate* this order, i.e., to 'unfold' it in the form of a new visual image.

A more striking example of implicate order can be demonstrated in the laboratory, with a transparent container full of a very viscous fluid, such as treacle, and equipped with a mechanical rotator that can 'stir' the fluid very slowly but very thoroughly. If an insoluble droplet of ink is placed in the fluid and the stirring device is set in motion, the ink drop is gradually transformed into a thread that extends over the whole fluid. The latter now appears to be distributed more or less at 'random' so that it is seen as some shade of grey. But if the mechanical stirring device is now turned in the opposite direction, the transformation is reversed, and the droplet of dye suddenly appears, reconstituted. (This illustration of the implicate order is discussed further in chapter 7.)

When the dye was distributed in what appeared to be a random way, it nevertheless had *some kind* of order which is different, for example, from that arising from another droplet originally placed in a different position. But this order is *enfolded* or *implicated* in the 'grey mass' that is visible in the fluid. Indeed, one could thus 'enfold' a whole picture. Different pictures would look indistinguishable and yet have different implicate orders, which differences would be revealed when they were explicated, as the stirring device was turned in a reverse direction.

What happens here is evidently similar in certain crucial ways to what happens with the hologram. To be sure there are differences. Thus, in a fine enough analysis, one could see that the *parts* of the ink droplet remain in a one-to-one correspondence as they are stirred up and the fluid moves continuously. On the other hand, in the functioning of the hologram there is no such one-to-one correspondence. So in the hologram (as also in experiments in a 'quantum' context), there is no way ultimately to reduce the implicate order to a finer and more complex type of explicate order.

All this calls attention to the relevance of a new distinction between implicate and explicate order. Generally speaking, the laws of physics have thus far referred mainly to the explicate order. Indeed, it may be said that the principle function of Cartesian coordinates is just to give a clear and precise description of explicate order. Now, we are proposing that in the formulation of the laws of physics, primary relevance is to be given to the implicate order, while the explicate order is to have a secondary kind of significance (e.g., as happened with Aristotle's notion of movement, after the development of classical physics). Thus, it may be expected that a description in terms of Cartesian coordinates can no longer be given a primary emphasis, and that a new kind of description will indeed have to be developed for discussing the laws of physics.

4 THE HOLOMOVEMENT AND ITS ASPECTS

To indicate a new kind of description appropriate for giving primary relevance to implicate order, let us consider once again the key feature of the functioning of the hologram, i.e., in each region of space, the order of a whole illuminated structure is 'enfolded' and 'carried' in the movement of light. Something similar happens with a signal that modulates a radio wave (see Figure 6.6). In all cases, the content or meaning that is 'enfolded' and 'carried' is primarily an order and a measure, permitting the development of a structure. With the radio wave, this structure can be that of a verbal communication, a visual image, etc., but with the hologram far more subtle structures can be involved in this way (notably three-dimensional structures, visible from many points of view).

More generally, such order and measure can be 'enfolded' and

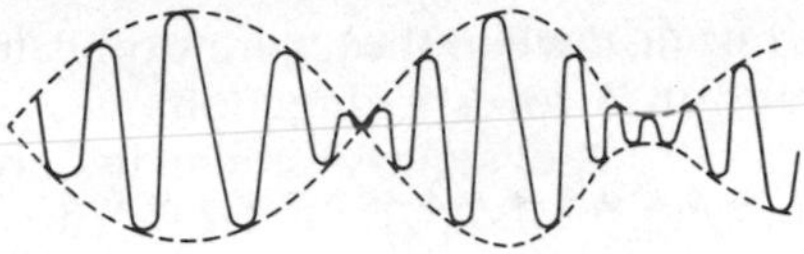

Figure 6.6

'carried' not only in electromagnetic waves but also in other ways (by electron beams, sound, and in other countless forms of movement). To generalize so as to emphasize undivided wholeness, we shall say that what 'carries' an implicate order is *the holomovement*, which is an unbroken and undivided totality. In certain cases, we can abstract particular aspects of the holomovement (e.g., light, electrons, sound, etc.), but more generally, all forms of the holomovement merge and are inseparable. Thus, in its totality, the holomovement is not limited in any specifiable way at all. It is not required to conform to any particular order, or to be bounded by any particular measure. Thus, *the holomovement is undefinable and immeasurable*.

To give primary significance to the undefinable and immeasurable holomovement implies that it has no meaning to talk of a *fundamental* theory, on which *all* of physics could find a *permanent* basis, or to which *all* the phenomena of physics could ultimately be reduced. Rather, each theory will abstract a certain aspect that is *relevant* only in some limited context, which is indicated by some appropriate measure.

In discussing how attention is to be called to such aspects, it is useful to recall that the word 'relevant' is a form obtained from the verb 'to relevate' which has dropped out of common usage, and which means 'to lift up' (as in 'elevate'). We can thus say in a particular context that may be under consideration, the general modes of description that belong to a given theory serve to *relevate* a certain content, i.e., to lift it into attention so that it stands out 'in relief'. If this content is pertinent in the context under discussion, it is said to be *relevant*, and otherwise, *irrelevant*.

To illustrate what it means to relevate certain aspects of the implicate order in the holomovement, it is useful to consider once again the example of the mechanical device for stirring a viscous fluid, as described in the previous section. Suppose that we first put in a droplet of dye and turn the stirring mechanism n times. We could then place another droplet of dye nearby and stir once again through n turns. We could repeat this process indefinitely, with a long series of droplets, arranged more or less along a line, as shown in Figure 6.7.

●●●●●●●●●●●●●●●●●●

Figure 6.7

Suppose, then, that after thus 'enfolding' a large number of droplets, we turn the stirring device in a reverse direction, but so rapidly that the individual droplets are not resolved in perception. Then we will see what appears to be a 'solid' object (e.g. a particle) moving continuously through space. This form of a moving object appears in immediate perception primarily because the eye is not sensitive to concentrations of dye lower than a certain minimum, so that one does not directly see the 'whole movement' of the dye. Rather, such perception *relevates a certain aspect*. That is to say, it makes this aspect stand out 'in relief' while the rest of the fluid is seen only as a 'grey' background' within which the related 'object' seems to be moving.

Of course, such an aspect has little interest *in itself*, i.e. apart from its *broader meaning*. Thus, in the present example, one possible meaning is that there *actually is* an autonomous object moving through the fluid. This would signify, of course, that the whole order of movement is to be regarded as similar to that in the immediately perceived aspect. In some contexts, such a meaning is pertinent and adequate (e.g., if we are dealing in the ordinary level of experience with a rock flying through the air). However, in the present context, a very different meaning is indicated, and this can be communicated only through a very different kind of description.

Such a description has to start by *conceptually* relevating certain broader orders of movement, going beyond any that are similar to those relevated in immediate perception. In doing this, one always begins with the holomovement, and then one abstracts special aspects which involve a totality broad enough for a proper description in the context under discussion. In the present example, this totality should include the whole movement of the fluid and the dye as determined by the mechanical stirring device, and the movement of the light, which enables us visually to perceive what is happening, along with the movement of the eye and nervous system, which determines the distinctions that can be perceived in the movement of light.

It may then be said that the content relevated in immediate

perception (i.e., the 'moving object') is a kind of *intersection* between two orders. One of these is the order of movement that brings about the possibility of a direct perceptual contact (in this case, that of the light and the response of the nervous system to this light), and the other is an order of movement that determines the detailed content that is perceived (in this case, the order of movement of the dye in the fluid). Such a description in terms of intersection of orders is evidently very generally applicable.[3]

It has already been seen that, in general, the movement of *light* is to be described in terms of 'the enfolding and carrying' of implicate orders that are relevant to a whole structure, in which analysis into separate and autonomous parts is not applicable (though, of course, in certain limited contexts, a description in terms of explicate orders will be adequate). In the present example, however, it is also appropriate to describe the movement of *the dye* in similar terms. That is to say, in the movement, certain implicate orders (in the distribution of dye) become explicate, while explicate orders become implicate.

To specify this movement in more detail, it is useful here to introduce a new *measure*, i.e., an 'implication parameter', denoted by T. In the fluid, this would be the number of turns needed to bring a given droplet of dye into explicate form. The total structure of dye present at any moment can then be regarded as a ordered series of substructures, each corresponding to a single droplet N with its implication parameter T_N.

Evidently, we have here a new notion of structure, for we no longer build structures solely as ordered and measured arrangements on which we join separate things, all of which are explicate together. Rather, we can now consider structures in which aspects of different degrees of implication (as measured by T) can be arranged in a certain order.

Such aspects can be quite complex. For example, we could implicate a 'whole picture' by turning the stirring device n times. We could then implicate a slightly different picture, and so on idefinitely. If the stirring device were turned rapidly in the reverse direction, we could see a 'three-dimensional scene' apparently consisting of a 'whole system' of objects in continuous movement and interaction.

In this movement, the 'picture' present at any given moment would consist only of aspects that can be explicated together (i.e., aspects corresponding to a certain value of the implication parameter T). As events happening at the same time are said to be *synchronous*, so aspects that can be explicated together can be called *synordinate*, while those that cannot be explicated together may then be called *asynordinate*. Evidently, the new notions of structure under discussion here involve *asynordinate* aspects, whereas previous notions involve only *synordinate* aspects.

It has to be emphasized here that the order of implication, as measured by the parameter T, has no necessary relationship to the order of time (as measured by *another* parameter, t). These two parameters are only related in a *contingent* manner (in this case by the rate of turning of the stirring device). It is the T parameter that is directly relevant to the description of the implicate structure, and not the t parameter.

When a structure is *asynordinate* (that is, constituted of aspects with different degrees of implication), then evidently the time order is not in general the primary one that is pertinent for the expression of law. Rather, as one can see by considering the previous examples, the *whole implicate order* is present at any moment, in such a way that the entire structure growing out of this implicate order can be described without giving any primary role to time. The law of the structure will then just be a law relating aspects with various degrees of implication. Such a law will, of course, not be deterministic *in time*. But, as has been indicated in chapter 5 determinism in time is not the only form of ratio or reason; and as long as we can find ratio or reason in the orders that are primarily relevant, this is all that is needed for law.

One can see in the 'quantum context' a significant similarity to the orders of movement that have been described in terms of the simple examples discussed above. Thus, as shown in Figure 6.8 'elementary particles' are generally observed by means of tracks that they are supposed to make in detecting devices (photographic emulsions, bubble chambers, etc). Such a track is evidently to be regarded as no more than an *aspect* appearing in immediate perception (as was done with the moving sequence of droplets of dye indicated in Figure 6.7). To describe it as the track of a 'particle'

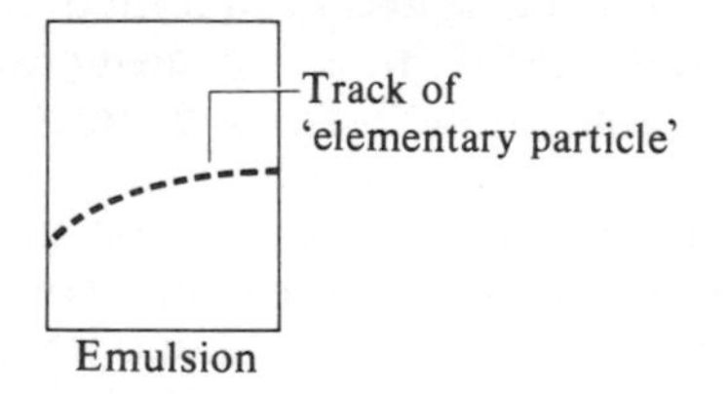

Figure 6.8

is then to assume in addition that the primarily relevant order of movement is similar to that in the immediately perceived aspect.

However, the whole discussion of the new order implicit in the quantum theory shows that such a description cannot coherently be maintained. For example, the need to describe movement discontinuously in terms of 'quantum jumps' implies that the notion of a well-defined orbit of a particle that connects the visible marks constituting the track cannot have any meaning. In any case, the wave-particle properties of matter show that the overall movement depends on the total experimental arrangement in a way that is not consistent with the idea of autonomous motion of localized particles; and, of course, the discussion of the Heisenberg microscope experiment indicates the relevance of a new order of undivided wholeness in which it has no meaning to talk about an observed object as if it were separate from the entire experimental situation in which observation takes place. So the use of the descriptive term 'particle' in this 'quantum' context is very misleading.

Evidently, we have here to deal with something that is similar in certain important ways to the example of stirring a dye into a viscous fluid. In both cases, there appears in immediate perception an explicate order that cannot consistently be regarded as autonomous. In the example of the dye, the explicate order is determined as an intersection of the implicate order of 'the whole movement' of the fluid and an implicate order of distinctions of density of dye that are relevated in sense perception. In the 'quantum' context, there similarly will be an intersection of an implicate order of some 'whole movement' corresponding to what we have called, for example, 'the electron', and another implicate order of distinctions that are relevated (and recorded) by our instruments. Thus, the word 'electron' should be regarded as no more than a name by which we call attention to a certain aspect of the holomovement, an aspect that can be discussed only by taking into account the entire experimental situation and that cannot be specified in terms of localized objects moving autonomously through space. And, of course, every kind of 'particle' which in current physics is said to be a basic constituent of matter will have to be discussed in the same sort of terms (so that such 'particles' are no longer considered as autonomous and separately existent). Thus, we come to a new general physical description in which 'everything implicates everything' in an order of undivided wholeness.

A mathematical discussion of how the 'quantum' context can

be assimilated in terms of the sort of implicate order discussed above is given in the appendix to this chapter.

5 LAW IN THE HOLOMOVEMENT

We have seen that in the 'quantum' context, the order in every immediately perceptible aspect of the world is to be regarded as coming out of a more comprehensive implicate order, in which all aspects ultimately merge in the undefinable and immeasurable holomovement. How, then, are we to understand the fact that descriptions involving the analysis of the world into autonomous components do actually work, at least in certain contexts (e.g., those in which classical physics is valid)?

To answer the question, we first note that the word 'autonomy' is based on two Greek words: 'auto', meaning 'self', and 'nomos' meaning 'law'. So, to be autonomous is to be *self-ruling*.

Evidently, nothing is 'a law unto itself'. At most, something may behave with a *relative and limited degree* of autonomy, under certain conditions and in certain degrees of approximation. Indeed, at the very least, each relatively autonomous thing (e.g., a particle) is limited by other such relatively autonomous things. Such a limitation is currently described in terms of *interaction*. However, we shall introduce here the word 'heteronomy' to call attention to a law in which many relatively autonomous things are related in this way, i.e., externally and more or less mechanically.

Now, what is characteristic of heteronomy is the applicability of *analytic descriptions*. (As pointed out in chapter 5, the root of the word 'analysis' is the Greek 'lysis' meaning 'to dissolve' or 'to loosen'. Since the prefix 'ana' means 'above', it may be said that 'to analyse' is to 'loosen from above', i.e., to obtain a broad view as if from a great height in terms of components that are regarded as autonomous and separately evident though in mutual interaction.)

As has been seen, however, in sufficiently broad contexts such analytic descriptions cease to be adequate. What is then called for is *holonomy*, i.e., the law of the whole. Holonomy does not totally deny the relevance of analysis in the sense discussed above. Indeed, 'the law of the whole' will generally include the possibility of describing the 'loosening' of aspects from each other, so that they will be relatively autonomous in limited contexts (as well as the possibility of describing the interactions of these aspects in a system of heteronomy). However, any form of relative autonomy (and heteronomy) is ultimately limited by holonomy, so that in a broad enough context such forms are seen to be merely aspects, relevated in the holomovement, rather than disjoint and separately existent things in interaction.

Scientific investigations have generally tended to begin by relevating apparently autonomous aspects of the totality. The study of the laws of these aspects has generally been emphasized at first, but as a rule this kind of study has led gradually to an awarenes that such aspects are related to others originally thought to have no significant bearing on the subject of primary interest.

From time to time, a wide range of aspects has been comprehended within a 'new whole'. But of course the general tendency until now has been to fix on this 'new whole' as a finally valid general order that is henceforth to be adapted (in the manner discussed in section 1) to fit any further facts that may be observed or discovered.

It is implied here, however, that even such a 'new whole' will itself be revealed as an aspect in yet another new whole. Thus, holonomy is not to be regarded as a fixed and final goal of scientific research, but rather as a movement in which 'new wholes' are continually emerging. And of course this further implies that the total law of the undefinable and immeasurable holomovement could never be known or specified or put into words. Rather, such a law has necessarily to be regarded as *implicit*.

The general question of the assimilation of the overall fact in physics in such a notion of law will now be discussed.

APPENDIX: IMPLICATE AND EXPLICATE ORDER IN PHYSICAL LAW

A.1 Introduction

In this appendix, the notions of implicate and explicate order that have been introduced earlier will be put into a more mathematical form.

It is important to emphasize, however, that mathematics and physics are not being regarded here as separate but mutually related structures (so that, for example, one could be said to apply mathematics to phsyics as paint is applied to wood). Rather, it is being suggested that mathematics and physics are to be considered as aspects of a single undivided whole.

In discussing this whole, we begin with the general language which is used for description in physics. We may then be said to *mathematize* this language, i.e. to articulate or define it in more detail so that it allows statements of greater precision from which a broad range of significant inferences may be drawn in a clear and coherent way.

In order that the general language and its mathematization shall be able to work together coherently and harmoniously, these two aspects have to be similar to each other in certain key ways, though they will, of course, be different in other ways (notably in that the mathematical aspect has greater possibilities for precision of inferences). Through a consideration of these similarities and differences, there can arise what may be called a sort of 'dialogue' in which new meanings common to both aspects are created. It is in this 'dialogue' that the wholeness of the general language and its mathematics is to be seen.

In this appendix we shall then indicate, though only in a very preliminary and provisional way, how we can mathematize the general language for developing implicate and explicate orders in a coherent and harmonious manner.

A.2 Euclidean Systems of Order and Measure

We begin with the mathematical description of explicate order.

Now, explicate order arises primarily as a certain aspect of sense perception and of experience with the content of such sense perception. It may be added that, in physics, explicate order generally reveals itself in the sensibly observable results of functioning of an instrument.

What is common to the functioning of instruments generally used in physical research is that the sensibly perceptible content is ultimately describable in terms of a Euclidean system of order and measure, i.e., one that can adequately be understood in terms of ordinary Euclidean geometry. We shall therefore begin with a discussion of Euclidean systems of order and measure.

In this discussion, we shall adopt the well-known view of the mathematician Klein, who considers the general transformations to be the essential determining features of a geometry. Thus, in a Euclidean space of three dimensions, there are three displacement operators D_i. Each of these operators defines a set of parallel lines which transform into themselves under the operation in question. Then, there are three rotation operators R_i. Each of these defines a set of concentric cylinders around the origin which transform into themselves under the operation in question. Together, they define concentric spheres which transform into themselves under the whole set of R_i. Finally, there is the dilatation operator

R_0, which transforms a sphere of a given radius into one of a different radius. Under this operation, the radial lines through the origin transform into themselves.

From any one set of operators R_i, R_0, we obtain another set R'_i, R'_0, corresponding to a different centre, by means of a displacement

$$(R'_i, R'_0) = D_j(R_i, R_0)\, D_j^{-1}.$$

From the D_i, we obtain a set of displacements D'_i in new directions by the rotation

$$D'_i = R_j D_i R_j^{-1}.$$

Now, if D_i is a certain displacement, $(D_i)^n$ will be a displacement of n similar steps. This means that displacements can be ordered naturally in an order similar to that of the integers. So we may describe displacements on a *numerical* scale. This gives not only an *order*, but also a *measure* (in so far as we treat successive displacements as equivalent in size).

Similarly, each rotation R_i determines an ordered and measured series $(R_i)^n$ of rotations, while a dilation R_0 determines an ordered and measured series $(R_0)^n$ of dilations.

It is clear that operations of this kind determine what is meant by parallelism and perpendicularity, as well as what is meant by congruence and similarity of geometric figures. Thus, they determine the essential feature of a Euclidean geometry, with its whole system of order and measure. It has to be kept in mind, however, that the whole set of operations is what is being taken as primarily relevant, while static elements (e.g., straight lines, circles, triangles, etc.) are now being regarded as 'invariant subspaces' of the operations and as configurations formed from these subspaces.

A3 Transformation and Metamorphosis

We now discuss the mathematical description of implicate order. Implicate order is generally to be described not in terms of simple geometric transformations, such as translations, rotations, and dilations, but rather in terms of a different kind of operation. In the interests of clarity, we shall therefore reserve the word *transformation* to describe a simple geometric change *within* a given explicate order. What happens in the broader context of implicate order we shall then call a *metamorphosis*. This word indicates that the change is much more radical than the change of position of orientation of a rigid body, and that it is in certain ways more like the changes from caterpillar to butterfly (in which everything alters in a thorough going manner while some subtle and highly implicit features remain invariant). Evidently, the change between an illuminated object and its hologram (or between an ink droplet and the 'grey mass' obtained by stirring it) is to be described as a metamorphosis rather than as a transformation.

We shall use the symbol M for a metamorphosis and T for a transformation, while E denotes a whole set of transformations that are relevant in a given explicate order (D_i, R_i, R_0). Under a metamorphosis, the set E will change into another set E given by

$$E' = MEM^{-1}.$$

This has hitherto generally been called a similarity transformation but from now on it will be called a similarity metamorphosis.

To indicate the essential features of a similarity metamorphosis, let us consider the example of the hologram. In this case, the appropriate metamorphosis M is determed by the Green's function relating amplitudes at the illuminated structure to those at the photographic plate. For waves of definite frequency ω the Green's function is

$$G(\mathbf{x} - \mathbf{y}) \simeq \{\exp[i(\omega/c)|\mathbf{x} - \mathbf{y}|]\}/|\mathbf{x} - \mathbf{y}|$$

where $\mathbf{x}$ is a coordinate relevant to the illuminated structure and $\mathbf{y}$ is one relevant to the plate. Thus, if $A(\mathbf{x})$ is the amplitude of the wave at the illuminated structure, then the amplitude $B(\mathbf{y})$ at the plate is

$$B(\mathbf{y}) \simeq \int (\{\exp[i(\omega/c)|\mathbf{x} - \mathbf{y}|]\}/|\mathbf{x} - \mathbf{y}|) A(\mathbf{x})\, d\mathbf{x}.$$

The entire illuminated structure is seen from the above equation to be 'carried' and 'enfolded' in each region of the plate in a way that evidently cannot be described in terms of a point-to-point transformation or correspondence between $\mathbf{x}$ and $\mathbf{y}$. The matrix $M(\mathbf{x}, \mathbf{y})$, which is essentially $G(\mathbf{x} - \mathbf{y})$, can thus be called a metamorphosis of the amplitudes at the illuminated structure into the amplitudes at the hologram.

Let us now consider the relationship between transformation E in the illuminated structure and concomitant changes in the hologram which follow these transformations. In the illuminated structure, E can be characterized as a point-to-point correspondence in which any similar locality is transformed into a similar locality. The corresponding change in the hologram is described by $E' = MEM^{-1}$. This is not a correspondence of points *in the hologram* to each other in which the property of locality of such sets of points would be preserved. Rather, each region of the hologram is changed in a way that depends on all other such regions. Nevertheless, the change E' in the hologram evidently determines the change E in the structure that can be seen when the hologram is illuminated with laser light.

Likewise, in a quantum context a unitary transformation (e.g., as given by a Green's functon operating on the state vector) can be understood as a metamorphosis in which point-to-point transformations of space and time that preserve locality are 'enfolded' into more general operations that are *similar* in the sense defined above and which nevertheless are not locality-preserving point-to-point transformations.

A4 Mathematization of the Description of Implicate Order

The next step is to discuss the mathematization of the language for the description of implicate order.

We begin by considering a metamorphosis M. By applying M again and again, we obtain $(M)^n$, which describes the enfolding of a given structure n times. If we then write $Q_n = (M)^n$, we have

$$Q_n : Q_{n-1} = Q_{n-1} : Q_{n-2} = M.$$

Thus, there is a series of similar differences in the Q_n (indeed, the differences are not only similar but are also all equal to M). As pointed out in chapter 5, such a series of similar differences indicates an *order*. Since the differences are in the degree of implication, this order is an implicate order. Moreover, in so far as successive operations M are regarded as equivalent, there is also a *measure*, in which n can be taken as an implication parameter.

If we think of the example of droplets of insoluble dye stirred into a viscous fluid (so that we let M describe the change of the droplet when the system is enfolded by a certain number of turns), then M^n describes the change of the droplet when subjected to n enfoldings. Each droplet is, however, inserted in a position that is displaced by a certain amount relative to the proceeding droplet. Let this displacement be denoted by D. The nth droplet first suffers the displacement D^n and then the metamorphosis is M^n, so that the net result is given by M^nD^n. Let us further suppose that the density of dye injected with each droplet can vary, and denote that of the dye injected into the *n*th droplet with the aid of the operation $Q_n = C_n M^n D^n$. The operator corresponding to the entire series of droplets is obtained by adding the contributions of each, to give

$$Q = \sum_n C_n M^n D^n.$$

Moreover, any number of structures, corresponding to Q, Q', Q'', etc., can also be superposed, to yield

$$R = Q + Q' + Q'' + \dots.$$

In addition, any such structure can itself undergo a displacement, such as D, and a metamorphosis, such as M, to yield

$$R' = MDR.$$

If the fluid were already a 'uniformly grey' background, we could give meaning to a negative coeffcient C_n as signifying the *removal* of a certain amount of dye from a region corresponding to a droplet (rather than to the addition of such dye to the region).

In the above discussion, each mathematical symbol corresponds to an operation (transformation and/or metamorphosis). There is a meaning to adding operations, to multiplying the result by a number C, and to multiplying operations by each other. If we further introduce a unit operation (one which leaves all operations unaltered in multiplication) and a zero operation (one which leaves all operations unaltered when added), we will have satisfied all the conditions needed for an algebra.

We see, then, that an algebra contains key features which are similar to the key features of structures built on implicate orders. Such an algebra thus makes possible a *relevant mathematization* that can be coherently related to the general language for discussing implicate orders.

Now, in the quantum theory an algebra similar to the one described above also plays a key role. Indeed, the theory is expressed in terms of linear operators (including a unit operator and a zero operator) that can be added to each other, multiplied by numbers, and multiplied by each other. All the content of the quantum theory can thus be put in terms of such an algebra.

Of course, in the quantum theory, the algebraic terms are interpreted as standing for 'physical observables' to which they correspond. However, in the approach that is being suggested here, such terms are not to be regarded as standing for anything in particular. Rather, they are to be considered as extensions of the general language. A single algebraic symbol is thus similar to a word, in the sense that its implict meaning comes out fully only in the way in which the language as a whole is used.

This approach is indeed used in a great deal of modern mathematics,[4] especially in number theory. Thus, one can start with what are called *undefinable symbols*. The meaning of such a symbol is never directly relevant. Rather, only relationships and operations in which these symbols take part are relevant.

What we are proposing here is that as we mathematize language in the way indicated above, there will arise orders, measures, and structures within the language which are similar to (but also different from) orders, measures, and structures that are to be perceived in common experience and in experience with the functioning of scientific instruments. As further indicated above, there can be a relationship between these two kinds of orders, measures, and structures, so that what we talk about and think about will have a common ratio or reason with what we can observe and do (see chapter 5 for a discussion of this sense of 'ratio' or 'reason').

This means, of course, that we do not regard terms like 'particle', 'charge', 'mass', 'position', 'momentum', etc., as having primary relevance in the algebraic language. Rather, at best, they will have to come out as high-level abstractions. As pointed out in this section, the real meaning of the 'quantum algebra' will then be that it is a mathematization of the general language, which enriches the latter and makes possible a more precisely articulated discussion of implicate order than is possible in terms of the general language alone.

Of course, algebra is in itself a limited form of mathematization. There is no reason in principle why we should not ultimately go on to other sorts of mathematization (involving, for example, rings and lattices or still more general structures which have yet to be created). However, it will be seen in this appendix that even within the limits of an algebraic structure, one can assimilate a very wide range of aspects of modern physics, and one can open up a great many interesting new avenues for exploration. It is therefore useful to go into the algebraic mathematization of the common language in some detail before going into more general kinds of mathematization.

A.5 Algebra and the Holomovement

We begin our exploration of the algebraic mathematization of the general language by calling attention to the fact that the primary meaning of an algebraic symbol is that it describes a certain kind of movement.

Thus, consider the set of undefinable algebraic terms denoted by A. It is characteristic of an algebra that these terms have a relationship given by

$$A_i A_j = \sum_k \lambda^K_{ij} A_K$$

where λ^K_{ij} is a set of numerical constants. This relationship means that when a given term A_i preceded another one A_j, the result is equivalent to a 'weighted sum' or superposition of terms (so that an algebra contains a sort of 'superposition principle' similar in key ways to that which holds in the quantum theory). In effect, one can say that although the term A_i is 'in itself' undefinable, it nevertheless signifies a certain sort of 'movement' of the total set of terms, in which each symbol A_j is replaced by (or changes into) a superposition of symbols $\Sigma\lambda^K_{ij}A_K$.

As pointed out earlier, however, in the general language for the description of implicate order the undefinable and immeasurable holomovement is considered as the totality in which all that is to be discussed is ultimately to be relevated. Similarly, in the algebraic mathematization of this general language, we consider as a totality an undefinable algebra in which the primary meaning of each term is that it signifies a 'whole movement' in all the terms of the algebra. Through this key similarity there arises the possibility of a coherent mathematization of the sort of general description that takes the totality to be the undefinable and immeasurable holomovement.

We can now go further along these lines. Thus, just as in the general language, we can consider relatively autonomous aspects of the holomovement, so in its mathematization, we can consider relatively autonomous sub-algebras which are aspects of the undefinable 'whole algebra'. As each aspect of the holomovement is ultimately limited in its autonomy by the law of the whole (i.e., the holonomy), so each sub-algebra is ultimately limited by the fact that the relevant law involves movements going outside those that can be described in terms of the sub-algebra in question.

A given physical context will then be describable in terms of an appropriate sub-algebra. As we approach the limits of this context, we will discover that such a description is inadequate and we will consider broader algebras until we find a discription that is adequate to the new context to which we have thus been led.

In the context of classical physics, for example, it is possible to abstract a sub-algebra corresponding to a set of Euclidean operations E. However, in a 'quantum' context, the 'law of the whole' involves metamorphoses M which lead out of this sub-algebra and into different (but similar) sub-algebras given by

$$E' = MEM^{-1}.$$

As pointed out, there are now indications that even the 'quantum' algebra is inadequate in yet broader contexts. So it is natural to go on to consider still broader algebras (and ultimately, of course, yet more general sorts of mathematization that may prove to be relevant).

A.6 Extension of Principle of Relativity to Implicate Orders

As a step into the inquiry into more comprehensive forms of mathematization, we shall point out the possibility of a certain extension of the principle of relativity to implicate orders that is suggested by considering how the quantum algebra limits the autonomy of the classical algebra in the way described above.

Now, in a classical context, any structure can be specified in terms of a set of operations E_1, E_2, E_3, . . . (which describe lengths, angles, congruence, similarity, etc.). When we go to a broader, 'quantum' content, we can arrive at similar operations, $E' = MEM^{-1}$. What this similarity means is that if any two elements, say E_1 and E_2, are related in a certain way in the description of a specified structure, then there is a set of elements E_1' and E_2' describing non-local 'enfolded' transformations that are related in a similar way. Or, to put it more concisely,

$$E_1 : E_2 :: E_1' : E_2'.$$

From this, it follows that if we are given a Euclidean system of order and measure with certain structures that are built on it, we can always obtain another system E' enfolded to relative E, and yet capable of having similar structures built on it.

Hitherto, the principle of relativity has taken a form which may be put as follows: 'Given any structural relationship as described in a frame of coordinates corresponding to a certain velocity, it is always possible to have a similar structural relationship as described in a frame of coordinates corresponding to any other velocity.' It follows from the discussion above, however, that the mathematization of the general language in terms of a 'quantum' algebra opens up the possibility of an extension of the prinicple of relativity. Such an extension is evidently similar to the principle of complementarity, in that when conditions are such that a given order corresponding to a set of operations E is explicate, then another order corresponding to similar operations $E' = MEM^{-1}$ is implicate (so that in a certain sense both orders cannot be defined together). However, it is different from the principle of complementarity in that the primary emphasis is now on orders and measures that are relevant to geometry, rather than on mutually incompatible experimental arrangements.

It follows from this extension of the principle of relativity that the idea of space as constituted of a set of unique and well-defined points, related topologically by a set of neighbourhoods and metrically by a definition of distance, is no longer adequate. Indeed, each set of Euclidean operations E' defines such a set of points, neighbourhoods, measures, etc., which are implicate relative to those defined by another set E'. The notion of space as a set of points with a topology and a metric is thus merely an aspect of a broader totality.

It will be helpful here to introduce a further new usage of language. In topology one can describe a space as covered by a *complex*, constituted of elementary figures (e.g. triangles or other basic polygonal cell forms), each of which is called a simplex. The word 'plex' is a form of the Latin 'plicare', which, as we have already seen earlier, means 'to fold'. So, 'simplex' means 'one-fold' and 'complex' means 'folded together', but in the sense of many separate objects that are joined to each other.

To describe the enfolding of an unlimited set of Euclidean systems of orders and measures into each other, we may then introduce the word *multiplex* (which is new in this context). This means 'many complexes all folded together'. Literally, this is also what is meant by 'manifold'. However, by custom, this last word has come to mean 'continuum'. So we are led to use the word multiplex to call attention to the primary relevance of implicate order and to the inadequacy of a description in terms of a continuum.

Thus far, space has generally been considered as a continuum that can be covered by a complex (which is evidently a form of explicate ordering of the space). Such a complex can be discussed in terms of coordinate systems. Thus, each simplex can be described with the aid of a locally Euclidean frame, and the whole space can then be treated through the use of a very large number of overlapping coordinate 'patches'. Or, alternatively, one may find a single set of curvilinear coordinates that is applicable over the entire space. The principle of relativity then states that all such coordinate systems furnish equivalent frames of description (i.e. equivalent for the expression of ratio, or reason, or law).

We can now go on to consider similar sets of operations E and E' which are implicate relative to each other. As pointed out above, we are extending the principle of relativity by supposing that the orders defined through any two operations E and E' are equivalent in the sense that the 'law of the whole' is such that similar structures can be built on each order. To help make clear what is meant here, we note that the orders of movement that are directly perceivable to the senses are generally regarded as explicate, while other orders (such as, for example, those appropriate to the description of 'an electron' in a quantum context) are taken to be implicate. However, according to the extended principle of relativity, one can equally well take the 'electron' order as explicate and our sensual order as implicate. This is to put ourselves (metaphorically) in the situation of 'the electron' and then to understand the latter by assimilating oneself to it and it to onself.

This evidently means a thoroughgoing wholeness in our thinking. Or, as put earlier, 'All implicates all', even to the extent that 'we ourselves' are implicated together with 'all that we see and think about'. So we are present everywhere and at all times, though only implicately (that is, implicitly).

The same is true of every 'object'. It is only in certain special orders of description that such objects appear as explicate. The general law, i.e., holonomy, has to be expressed in all orders, in which all objects and all times are 'folded together'.

A.7 Some Preliminary Suggestions Concerning Law in a Multiplex

We shall now give a few preliminary suggestions as to the lines of inquiry into general law as formulated in terms of a multiplex rather than in terms of a continuum.

We begin by recalling that classical descriptions are relevant only in a context in which the expression of the law is limited to a particular sub-algebra corresponding to a given Euclidean system of order and measure. If this system is extended to time as well as space, then such a law can be compatible with special relativity.

The essential feature of special relativity is that the speed of light is an invariant limit for the propagation of signals (and causal influences). In this connection, we note that a signal will always be constituted of a certain explicate order of events, and that in a context in which this explicate order ceases to be relevant, the notion of signal will also cease to be relevant (e.g., if an order is 'enfolded' throughout all of space and time, it cannot coherently be regarded as constituting a signal that would propagate information from one place to another over a period of time). This means that where implicate order is involved, the descriptive language of special relativity will, in general, no longer be applicable.

The general theory of relativity is similar to the special theory, in that in each region of space–time there is a light cone which defines a limiting signal velocity. It is different, however, in that each region has its own local coordinate frame (denoted by m) related to those of its neighbours (denoted by n) through certain general linear transformations T_{mn}. But a local coordinate frame is, in our point of view, to be regarded as an expression of a corresponding Euclidean system of order and measure (which would, for example, generate the lines of the frame in question as invariant subspaces of the operations E). We therefore consider the Euclidean systems of operations E_m and E_n and the transformations relating them:

$$E_n = T_{mn} E_m T_{mn}^{-1}.$$

When we consider a series of transformations of these systems around a closed circuit of patches, we arrive at what is in mathematical terms called the 'holonomy group'. In one sense, this name is appropriate, for this group does determine the character of the 'whole space'. Thus, in general relativity, this group is equivalent to the Lorentz group, which is compatible with the requirement of an invariant 'local light cone'. The use of a different group here would of course imply a correspondingly different character to the 'whole space'.

In another sense, however, it would be better to consider the group in question as an 'autonomy group' rather than as a 'holonomy group', for, in general relativity (as well as in a wide class of modern field theories), the general law is invariant to arbitrary 'gauge transformations' of the frames in each region, $E_m' = R_m E_m R_m^{-2}$. The meaning of these transformations can be seen by considering several neighbouring regions, each containing a localized structure, i.e., one which has a negligible connection with neighbouring structures (so that one may appropriately regard the space between them as empty, or approximately so). The significance of gauge invariance is then that the laws are such that any two structures can be transformed independently of each other, at least within certain limits (e.g., as long as there is sufficient 'empty space' between them). An example of such relative autonomy of structures is that objects that are not too close can be turned and translated relative to each other. Evidently, it is this particular feature of 'law of the whole' (i.e., gauge invariance) which allows for relative autonomy of the kind described above.

As we go on to a quantum context, the 'law of the whole' (i.e., the generalization of what is meant by 'holonomy group' in Rieman geometry) will involve metamorphosis M as well as transformations T. This will bring us to the multiplex, in which new kinds of order and measure will be relevant.

It is important, however, to emphasize that the 'law of the whole' will not just be a transcription of current quantum theory to a new language. Rather, the entire context of physics (classical and quantum) will have to be assimilated in a different structure, in which space, time, matter, and movement are described in new ways. Such assimilation will then lead on to new avenues to be explored, which cannot even be thought about in terms of current theories.

We shall here indicate only a few of the many possibilities of this kind.

First, we recall that we begin with an undefinable total algebra and take out sub-algebras that are suitable for the description of certain contexts of physical research. Now, mathematicians have already worked out certain interesting and potentially relevant features of such sub-algebras.

Thus, consider a given sub-algebra A. Among its terms A_i, there may be some A_N which are nilpotent, i.e., which have the property that some powers of A_N (say $(A_N)^s$) are zero. Among these, there is a subset of terms A_p which are *properly nilpotent*, i.e. which remain nilpotent when multiplied by any term of the algebra A_i (so that $(A_iA_p)^s = 0$).

As an example, consider first a clifford algebra, in which every term is properly nilpotent. However, in a fermionic algebra, with terms C_i and C_j^*, each C_i and C_j^* is nilpotent (i.e., $(C_i)^2 = (C_j^*)^2 = 0$) but not properly nilpotent (i.e., $(C_i^* \; C_j)^2 \neq 0$).

One may say that properly nilpotent terms describe movements which ultimately lead to features that vanish. Thus, if we are seeking to describe invariant and relatively permanent features of movement, we should have an algebra that has no properly nilpotent terms. Such an algebra can always be obtained from any algebra A by subtracting the properly nilpotent terms to give what is called the *difference algebra*.

We now consider the following theorem.[5] Every different algebra can be expressed in terms of products of a matrix algebra (i.e., an algebra whose rules of multiplication are similar to those of matrices) and a division algebra (i.e.; an algebra in which the product of two non-zero terms is never zero).

As regards the division algebra, the possible types of these depend on the fields over which the numerical coefficients are taken. If this field is that of the real numbers, then there are exactly three division algebras, the real numbers themselves, an algebra of order two, which is equivalent to complex numbers, and the real quaternions. On the other hand, over the field of complex numbers, the only division algebra is that of the complex numbers themselves (this explains why quaternions, extended to include complex coefficients, become a two-rowed matrix algebra).

It is significant that by mathematizing the general language in terms of an initially undefined and unspecified algebra, we arrive naturally at the sort of algebras used in current quantum theory for 'particles with spin', i.e. products of matrices and quaternions. These algebras have in addition, however, a significance going beyond that of technical calculations carried out in the quantum theory. For example, the quaternions imply invariance under a group of transformations similar to rotations in three-dimensional space (which can be extended in a simple way to groups similar to the Lorentz group). This indicates that, in some sense, the key transformations determining the (3 + 1)-dimensional order of 'relativist space–time' are already contained in the holomovement, described through implicate order, mathematized in terms of algebra.

More precisely, it can be said that, starting from a general algebraic mathematization of the language and asking for those features which are relatively permanent or invariant (described by algebras without properly nilpotent terms) and those features which are not restricted to a particular scale (described by algebras whose terms can be multiplied by an arbitrary real number), we have arrived at transformations determining an order equivalent to that of relativistic space–time. This means, however, that if we considered impermanent and non-invariant features (implying algebras with properly nilpotent terms) and features that are restricted to particular scales (implying algebras over the rationals or over finite number fields), then entirely new orders (not reducible at all to (3 + 1)-dimensional order) may become relevant. It thus becomes clear that there is here a wide area for possible exploration.

A further area for exploration would be in the development of a new description combining classical and quantum aspects in a single or more comprehensive structure of language. Instead of regarding classical and quantum languages as separate but related by some sort of correspondence (as is generally done in current theories), one can, along the lines already indicated in this appendix, inquire into the possibility of abstracting these as limiting cases of languages mathematized in terms of broader algebras. To do this could evidently lead to different theories, having a new content, going beyond those of both classical and quantum theories. In this regard, it would be particularly interesting to see if algebraic structures would be discovered which lead also to relativistic notions as limiting cases (e.g., in terms of algebras over finite number fields, rather than over the reals). Such theories might be expected to be free of the infinities of current theories, and to lead to a generally coherent treatment of the problems that the current theories cannot solve.

7

The enfolding-unfolding universe and consciousness

1 INTRODUCTION

Throughout this book the central underlying theme has been the unbroken wholeness of the totality of existence as an undivided flowing movement without borders.

It seems clear from the discussion in the previous chapter that the implicate order is particularly suitable for the understanding of such unbroken wholeness in flowing movement, for in the implicate order the totality of existence is enfolded within each region of space (and time). So, whatever part, element, or aspect we may abstract in thought, this still enfolds the whole and is therefore intrinsically related to the totality from which it has been abstracted. Thus, wholeness permeates all that is being discussed, from the very outset.

In this chapter we shall give a non-technical presentation of the main features of the implicate order, first as it arises in physics, and then as it may be extended to the field of consciousness, to indicate certain general lines along which it is possible to comprehend both cosmos and consciousness as a single unbroken totality of movement.[1]

2 RÉSUMÉ, CONTRASTING MECHANISTIC ORDER IN PHYSICS WITH IMPLICATE ORDER

It will be helpful to begin by giving a résumé of some of the main points that have been made earlier, contrasting the generally accepted mechanistic order in physics and the implicate order.

Let us first consider the mechanistic order. As indicated in chapters 1 and 5, the principal feature of this order is that the world is regarded as constituted of entities which are *outside of each other*, in the sense that they exist independently in different regions of space (and time) and interact through forces that do not bring about any changes in their essential natures. The machine gives a typical illustration of such a system of order. Each part is formed (e.g., by stamping or casting) independently of the others, and interacts with the other parts only through some kind of external contact. By contrast, in a living organism, for example, each part grows in the context of the whole, so that it does not exist independently, nor can it be said that it merely 'interacts' with the others, without itself being essentially affected in this relationship.

As pointed out in chapter 1, physics has become almost totally committed to the notion that the order of the universe is basically mechanistic. The most common form of this notion is that the world is assumed to be constituted of a set of separately existent, indivisible and unchangeable 'elementary particles', which are the fundamental 'building blocks' of the entire universe. Originally, these were thought to be atoms, but atoms were eventually divided into electrons, protons and neutrons. These latter were thought to be the absolutely unchangeable and indivisible constituents of all matter, but then, these were in turn found to be subject to transformation into hundreds of different kinds of unstable particles, and now even smaller particles called 'quarks' and 'partons' have been postulated to explain these transformations. Though these have not yet been isolated there appears to be an unshakable faith among physicists that either such particles, or some other kind yet to be discovered, will eventually make possible a complete and coherent explanation of everything.

The theory of relativity was the first significant indication in physics of the need to question the mechanistic order. As explained in chapter 5, it implied that no coherent concept of an independently existent particle is possible, neither one in which the particle would be an extended body, nor one in which it would be a dimensionless point. Thus, a basic assumption underlying the generally accepted form of mechanism in physics has been shown to be untenable.

To meet this fundamental challenge, Einstein proposed that the particle concept no longer be taken as primary, and that instead reality be regarded from the very beginning as constituted of fields, obeying laws that are consistent with the requirements of the theory of relativity. A key new idea of this 'unified field theory' of Einstein is that the field equations be *non-linear*. As stated in chapter 5, these equations could have solutions in the form of localized pulses, consisting of a region of intense field that could move through space stably as a whole, and that could thus provide a model of the 'particle'. Such pulses do not end abruptly but spread out to arbitrarily large distances with decreasing intensity. Thus the field structures associated with two pulses will merge and flow together in one unbroken whole. Moreover, when two pulses come close together, the original particle-like forms will be so radically altered that there is no longer even a resemblance to a structure consisting of two particles. So, in terms of this notion, the idea of a separately and independently existent particle is seen to be, at best, an abstraction furnishing a valid approximation only in a certain limited domain. Ultimately, the entire universe (with all its 'particles', including those constituting human beings, their laboratories, observing instruments, etc.) has to be understood as a single undivided whole, in which analysis into separately and independently existent parts has no fundamental status.

As has been seen in chapter 5, however, Einstein was not able to obtain a generally coherent and satisfactory formulation of his unified field theory. Moreover (and perhaps more important in the context of our discussion of the mechanistic approach to physics) the field concept, which is his basic starting point, still retains the essential features of a mechanistic order, for the fundamental entities, the fields, are conceived as existing outside of each other, at separate points of space and time, and are assumed to be connected with each other only through external relationships which indeed are also taken to be local, in the sense that only those field elements that are separated by 'infinitesimal' distances can affect each other.[2]

Though the unified field theory was not successful in this attempt to provide an ultimate mechanistic basis for physics in terms of the field concept, it nevertheless did show in a concrete way how consistency with the theory of relativity may be achieved by deriving the particle concept as an abstraction from an unbroken and undivided totality of existence. Thus, it helped to strengthen the challenge posed by relativity theory to the prevailing mechanistic order.

The quantum theory presents, however, a much more serious challenge to this mechanistic order, going far beyond that provided by the theory of relativity. As seen in chapter 5, the key features of the quantum theory that challenge mechanism are:

1. Movement is in general *discontinuous*, in the sense that action is constituted of *indivisible quanta* (implying also that an electron, for example, can go from one state to another, without passing through any states in between).

2. Entities, such as electrons, can show different properties (e.g., particle-like, wavelike, or something in between), depending on the environmental context within which they exist and are subject to observation.

3. Two entities, such as electrons, which initially combine to form a molecule and then separate, show a peculiar non-local relationship, which can best be described as a non-causal connection of elements that are far apart[3] (as demonstrated in the experiment of Einstein, Podolsky and Rosen[4]).

It should be added of course that the laws of quantum mechanics are statistical and do not determine individual future events uniquely and precisely. This is, of course, different from classical laws, which do in principle determine these events. Such indeterminism is, however, not a serious challenge to a mechanistic order, i.e., one in which the fundamental elements are independently existent, lying outside each other, and connected only by external relationships. The fact that (as in a pinball machine) such elements are related by the rules of chance (expressed mathematically in terms of the theory of probability) does not change the basic externality of the elements[5] and so does not essentially affect the question of whether the fundamental order is mechanistic or not.

The three key features of the quantum theory given do, however, clearly show the inadequacy of mechanistic notions. Thus, if all actions are in the form of discrete quanta, the interactions between different entities (e.g., electrons) constitute a single structure of indivisible links, so that the entire universe has to be thought of as an unbroken whole. In this whole, each element that we can abstract in thought shows basic properties (wave or particle, etc.) that depend on its overall environment, in a way that is much more reminiscent of how the organs constituting living beings are related, than it is of how parts of a machine interact. Further, the non-local, non-causal nature of the relationships of elements distant from each other evidently violates the requirements of separateness and independence of fundamental constituents that is basic to any mechanistic approach.

It is instructive at this point to contrast the key features of relativistic and quantum theories. As we have seen, relativity theory requires continuity, strict causality (or determinism) and locality. On the other hand, quantum theory requires non-continuity, non-causality and non-locality. So the basic concepts of relativity and quantum theory directly contradict each other. It is therefore hardly surprising that these two theories have never been unified in a consistent way. Rather, it seems most likely that such a unification is not actually possible. What is very probably needed instead is a qualitatively new theory, from which both relativity and quantum theory are to be derived as abstractions, approximations and limiting cases.

The basic notions of this new theory evidently cannot be found by beginning with those features in which relativity and quantum theory stand in direct contradition. The best place to begin is with what they have basically in common. This is undivided wholeness. Though each comes to such wholeness in a different way, it is clear that it is this to which they are both fundamentally pointing.

To begin with undivided wholeness means, however, that we must drop the mechanistic order. But this order has been, for many centuries, basic to all thinking on physics. As brought out in chapter 5, the mechanistic order is most naturally and directly expressed through the Cartesian grid. Though physics has changed radically in many ways, the Cartesian grid (with minor modifications, such as the use of curvilinear coordinates) has remained the one key feature that has not changed. Evidently, it is not easy to change this, because our notions of order are pervasive, for not only do they involve our thinking but also our senses, our feelings, our intuitions, our physical movement, our relationships with other people and with society as a whole and, indeed, every phase of our lives. It is thus difficult to 'step back' from our old notions of order sufficiently to be able seriously to consider new notions of order.

To help make it easier to see what is meant by our proposal of new notions of order that are appropriate to undivided wholeness, it is therefore useful to start with examples that may directly involve sense perception, as well as with models and analogies that illustrate such notions in an imaginative and intuitive way. In chapter 6 we began by noting that the photographic lens is an instrument that has given us a very direct kind of sense perception of the meaning of the mechanistic order, for by bringing about an approximate correspondence between points on the object and points on the photographic image, it very strongly calls attention to the separate elements into which the object can be analysed. By making possible the point-to-point imaging and recording of things that are too small to be seen with the naked eye, too big, to fast, too slow, etc., it leads us to believe that eventually everything can be *perceived* in this way. From this grows the idea that there is nothing that cannot also be *conceived* as constituted of such localized elements. Thus, the mechanistic approach was greatly encouraged by the development of the photographic lens.

We then went on to consider a new instrument, called the *hologram*. As explained in chapter 6, this makes a photographic record of the interference pattern of light waves that have come off an object. The key new feature of this record is that each part contains information about the *whole object* (so that there is no point-to-point correspondence of object and recorded image). That is to say, the form and structure of the entire object may be said to be *enfolded* within each region of the photographic record. When one shines light on any region, this form and structure are then *unfolded*, to give a recognizable image of the whole object once again.

We proposed that a new notion of order is involved here, which we called the *implicate order* (from a Latin root meaning 'to enfold' or 'to fold inward'). In terms of the implicate order one may say that everything is enfolded into everything. This contrasts with the *explicate order* now dominant in physics in which things are *unfolded* in the sense that each thing lies only in its own particular region of space (and time) and outside the regions belonging to other things.

The value of the hologram in this context is that it may help to bring this new notion of order to our attention in a sensibly perceptible way; but of course, the hologram is only an instrument whose function is to make a static record (or 'snapshot') of this order. The actual order itself which has thus been recorded is in the complex movement of electromagnetic fields, in the form of light waves. Such movement of light waves is present everywhere and in principle enfolds the entire universe of space (and time) in each region (as can be demonstrated in any such region by placing one's eye or a telescope there, which will 'unfold' this content).

As pointed out in chapter 6, this enfoldment and unfoldment takes place not only in the movement of the electromagnetic field but also in that of other fields, such as the electronic, protonic, sound waves, etc. There is already a whole host of such fields that are known, and any number of additional ones, as yet unknown, that may be discovered later. Moreover, the movement is only approximated by the classical concept of fields (which is generally used for the explanation of how the hologram works). More accurately, these fields obey quantum-mechanical laws, implying the properties of discontinuity and non-locality, which we have already mentioned (and which we shall discuss again later in this chapter). As we shall see later, even the quantum laws may only be abstractions from still more general laws, of which only some outlines are now vaguely to be seen. So the totality of movement of enfoldment and unfoldment may go immensely beyond what has revealed itself to our observations thus far.

In chapter 6 we called this totality by the name *holomovement*. Our basic proposal was then that *what is* is the holomovement, and that everything is to be explained in terms of forms derived from this holomovement. Though the full set of laws governing its totality is unknown (and, indeed, probably unknowable) nevertheless these laws are assumed to be such that from them may be abstracted relatively autonomous or independent sub-totalities of movement (e.g., fields, particles, etc.) having a certain recurrence and stability of their basic patterns of order and measure. Such sub-totalities may then be investigated, each in its own right, without our having first to know the full laws of the holomovement. This implies, of course, that we are not to regard what we find in such investigations as having an absolute and final validity, but rather we have always to be ready to discover the limits of independence of any relatively autonomous structure of law, and from this to go on to look for new laws that may refer to yet larger relatively autonomous domains of this kind.

Up till now we have contrasted implicate and explicate orders, treating them as separate and distinct, but as suggested in chapter 6, the explicate order can be regarded as a particular or distinguished case of a more general set of implicate orders from which latter it can be derived. What distinguishes the explicate order is that what is thus derived is a set of recurrent and relatively stable elements that are *outside* of each other. This set of elements (e.g., fields and particles) then provides the explanation of that domain of experience in which the mechanistic order yields an adequate treatment. In the prevailing mechanistic approach, however, these elements, assumed to be separately and independently existent, are taken as constituting the basic reality. The task of science is then to start from such parts and to derive all wholes through abstraction, explaining them as the results of interactions of the parts. On the contrary, when one works in terms of the implicate order, one begins with the undivided wholeness of the universe, and the task of science is to derive the parts through abstraction from the whole, explaining them as approximately separable, stable and recurrent, but externally related elements making up relatively autonomous sub-totalities, which are to be described in terms of an explicate order.

3 THE IMPLICATE ORDER AND THE GENERAL STRUCTURE OF MATTER

We shall now go on to give a more detailed account of how the general structure of matter may be understood in terms of the implicate order. To do this we shall begin by considering once again the device discussed in chapter 6, which served as an analogy, illustrating certain essential features of the implicate order. (It must be emphasized, however, that it is *only* an analogy and that, as will be brought out in more detail later, its correspondence with the implicate order is limited.)

This device consisted of two concentric glass cylinders, with a highly viscous fluid such as glycerine between them, which is arranged in such a way that the outer cylinder can be turned very slowly, so that there is negligible diffusion of the viscous fluid. A droplet of insoluble ink is placed in the fluid, and the outer cylinder is then turned, with the result that the droplet is drawn out into a fine thread-like form that eventually becomes invisible. When the cylinder is turned in the opposite direction the thread-form draws back and suddenly becomes visible, as a droplet essentially the same as the one that was there originally.

It is worth while to reflect carefully on what is actually happening in the process described above. First, let us consider an element of fluid. The parts at larger radii will move faster than those at smaller radii. Such an element will therefore be deformed, and this explains why it is eventually drawn out into a long thread. Now, the ink droplet consists of an aggregate of carbon particles that are initially suspended in such an element of fluid. As the element is drawn out the ink particles will be carried with it. The set of particles will thus spread out over such a large volume that their density falls below the minimum threshold that is visible. When the movement is reversed, then (as is known from the physical laws governing viscous media) each part of the fluid retraces its path, so that eventually the thread-like fluid element draws back to its original form. As it does so, it carries the ink particles with it, so that eventually they, too, draw together and become dense enough to pass the threshold of perceptibility, so emerging once again as visible droplets.

When the ink particles have been drawn out into a long thread, one can say that they have been *enfolded* into the glycerine, as it might be said that an egg can be folded into a cake. Of course, the difference is that the droplet can be unfolded by reversing the motion of the fluid, while there is no way to unfold the egg (this is because the material here undergoes irreversible diffusive mixing).

The analogy of such enfoldment and unfoldment to the implicate order introduced in connection with the hologram is quite good. To develop this analogy further, let us consider two ink droplets close to each other, and to make visualization easier we will suppose that the ink particles in one droplet are red, while those in the other are blue. If the outer cylinder is then turned, each of the two separate elements of fluid in which the ink particles are suspended will be drawn out into a thread-like form, and the two thread-like forms will, while remaining separate and distinct, weave through each other in a complex pattern too fine to be perceptible to the eye (rather like the interference pattern that is recorded on the hologram, which has, however, quite a different origin). The ink particles in each droplet will of course be carried along by the fluid motions, but with each particle remaining in its own thread of fluid. Eventually, however, in any region that was large enough to be visible to the eye, red particles from the one droplet and blue particles from the other will be seen to intermingle, apparently at random. When the fluid motions are reversed, however, each thread-like element of fluid will draw back into itself until eventually the two gather into clearly separated regions once again. If one were able to watch what is happening more closely (e.g., with a microscope) one would see red and blue particles that were close to each other beginning to separate, while particles of a given colour that were far from each other would begin to come together. It is almost as if distant particles of a given colour had 'known' that they had a common destiny, separate from that of particles of the other colour, to which they were close.

Of course, there is in this case actually no such 'destiny'. Indeed, we have explained all that has happened mechanically, through the complex movements of the fluid elements in which the ink particles are suspended. But we have to recall here that this device is only an analogy, intended to illustrate a new notion of order. To allow this new notion to stand out clearly, it is necessary to begin by focusing our attention on the ink particles alone, and to set aside the consideration of the fluid in which they are suspended, at least for the moment. When the sets of ink particles from each droplet have been drawn out into an invisible thread, so that particles of both colours intermingle, one can nevertheless say that *as an ensemble* each set is, in a certain way, distinct from the other. This distinction is not in general evident to the senses, but it has a certain relationship to the total situation out of which the ensembles have come. This situation includes the glass cylinders, the viscous fluid and its movements, and the original distribution of ink particles. It may then be said that each ink particle belongs to a certain distinct ensemble and that it is bound up with the others in this ensemble by the force of an overall necessity, inherent in this total situation, which can bring the whole set to a common end (i.e., to reconstitute the form of a droplet).

In the case of this device, the overall necessity operates mechanically as the movement of fluid, according to certain well-known laws of hydrodynamics. As indicated earlier, however, we will eventually drop this mechanical analogy and go on to consider the holomovement. In the holomovement, there is still an overall necessity (which in chapter 6 we called 'holonomy') but its laws are no longer mechanical. Rather, as pointed out in section 2 of this chapter, its laws will be in a first approximation those of the quantum theory, while more accurately they will go beyond even these, in ways that are at present only vaguely discernible. Nevertheless, certain similar principles of distinction will prevail in the holomovement as in the analogy of the device made up of glass cylinders. That is to say, ensembles of elements which intermingle or inter-penetrate in space can nevertheless be distinguished, but only in the context of certain total situations, in which the members of each ensemble are related through the force of an overall necessity inherent in these situations, that can bring them together in a specifiable way.

Now that we have established a new kind of distinction of ensembles that are enfolded together in space, we can go on to put these distinctions into an *order*. The simplest notion of order is that of a sequence or succession. We shall start with such a simple idea and develop it later to much more complex and subtle notions of order.

As shown in chapter 5, the essence of a simple, sequential order is in the series of relationships among distinct elements:

$$A : B :: B : C :: C : D \ldots.$$

For example, if A represents one segment of a line, B the succeeding one, etc., the sequentiality of segments of the line follows from the above set of relationships.

Let us now return to our ink-in-fluid analogy, and suppose that we have inserted into the fluid a large number of droplets, set close to each other and arranged in a line (this time we do not suppose different colours). These we label as $A, B, C, D \ldots.$ We then turn the outer cylinder many times, so that each of the droplets gives rise to an ensemble of ink particles, enfolded in so large a region of space that particles from all the droplets intermingle. We label the successive ensembles $A', B', C', D' \ldots.$

It is clear that, in some sense, an entire linear order has been enfolded into the fluid. This order may be expressed through the relationships

$$A' : B' :: B' : C' :: C' : D' \ldots.$$

This order is not present to the senses. Yet its reality may be demonstrated by reversing the motion of the fluid, so that the ensembles, $A', B', C', D' \ldots$, will unfold to give rise to the original linearly arranged series of droplets, $A, B, C, D \ldots.$

In the above, we have taken a pre-existent explicate order, consisting of ensembles of ink particles arranged along a line, and transformed it into an order of enfolded ensembles, which is in some key way similar. We shall next consider a more subtle kind of order, not derivable from such a transformation.

Suppose now that we insert an ink droplet, A, and turn the outer cylinder n times. We then insert a second ink droplet, B, at the same place, and again turn the cylinder n times. We keep up this procedure with further droplets, $C, D, E \ldots.$ The resulting ensembles of ink particles, $a, b, c, d, e, \ldots$, will now differ in a new way, for, when the motion of the fluid is reversed, the ensembles will successively come together to form droplets in an order opposite to the one in which they were put in. For example, at a certain stage the particles of ensemble d will come together (after which they will be drawn out into a thread again). This will happen to those of c, then to b, etc. It is clear from this that ensemble d is related to c as c is to b, and so on. So these ensembles form a certain sequential order. However, this is in no sense a transformation of a linear order in space (as was that of the sequence A', B', C', D' . . . , that we considered earlier), for in general only one of these ensembles will unfold at a time; when any one is unfolded, the rest are still enfolded. In short, we have an order which cannot all be made explicate at once and which is nevertheless real, as may be revealed when successive droplets become visible as the cylinder is turned.

We call this an *intrinsically implicate order*, to distinguish it from an order that may be enfolded but which can unfold all at once into a single explicate order. So we have here an example of how, as stated in section 2, an explicate order is a particular case of a more general set of implicate orders.

Let us now go on to combine both of the above-described types of order.

We first insert a droplet A, in a certain position and turn the cylinder n times. We then insert a droplet, B, in a slightly different position and turn the cylinder n more times (so that A has been enfolded by $2n$ turns). We then insert C further along the line AB and turn n more times, so that A has been enfolded by $3n$ turns, B $2n$ turns, and C by n turns. We proceed in this way to enfold a large number of droplets. We then move the cylinder fairly rapidly in the reverse direction. If the rate of emergence of droplets is faster than the minimum time of resolution of the human eye, what we will see is apparently a particle moving continuously and crossing the space.

Such enfoldment and unfoldment in the implicate order may evidently provide a new model of, for example, an electron, which is quite different from that provided by the current mechanistic notion of a particle that exists at each moment only in a small region of space and that changes its position continuously with time. What is essential to this new model is that the electron is instead to be understood through a total set of enfolded ensembles, which are generally not localized in space. At any given moment one of these may be unfolded and therefore localized, but in the next moment, this one enfolds to be replaced by the one that follows. The notion of continuity of existence is approximated by that of very rapid recurrence of similar forms, changing in a simple and regular way (rather as a rapidly spinning bicycle wheel gives the impression of a solid disc, rather than of a sequence of rotating spokes). Of course, more fundamentally, the particle is only an abstraction that is manifest to our senses. *What is* is always a totality of ensembles, all present together, in an orderly series of stages of enfoldment and unfoldment, which intermingle and inter-penetrate each other in principle throughout the whole of space.

It is further evident that we could have enfolded any number of such 'electrons', whose forms would have intermingled and inter-penetrated in the implicate order. Nevertheless, as these forms unfolded and became manifest to our senses, they would have come out as a set of 'particles' clearly separated from each other. The arrangement of ensembles could have been such that these particle-like manifestations came out 'moving' independently in straight lines, or equally well, along curved paths that were mutually related and dependent, as if there had been a force of interaction between them. Since classical physics traditionally aims to explain everything in terms of interacting systems of particles, it is clear that in principle one could equally well treat the entire domain that is correctly covered by such classical concepts in terms of our model of ordered sequences of enfolding and unfolding ensembles.

What we are proposing here is that in the quantum domain this model is a great deal better than is the classical notion of an interacting set of particles. Thus, although successive localized manifestations of an electron, for example, may be very close to each other, so that they approximate a continuous track, this need not always be so. In principle, discontinuities may be allowed in the manifest tracks – and these may, of course, provide the basis of an explanation of how, as stated in section 2, an electron can go from one state to another without passing through states in between. This is possible, of course, because the 'particle' is only an abstraction of a much greater totality of structure. This abstraction is what is manifest to our senses (or instruments) but evidently there is no reason why it has to have continuous movement (or indeed continuous existence).

Next, if the total context of the process is changed, entirely new modes of manifestation may arise. Thus, returning to the ink-in-fluid analogy, if the cylinders are changed, or if obstacles are placed in the fluid, the form and order of manifestation will be different. Such a dependence – the dependence of what manifests to observation on the total situation – has a close parallel to a feature which we have also mentioned in section 2, i.e., that according to the quantum theory electrons may show properties resembling either those of particles or those of waves (or of something in between) in accordance with the total situation involved in which they exist and in which they may be observed experimentally.

What has been said thus far indicates that the implicate order gives generally a much more coherent account of the quantum properties of matter than does the traditional mechanistic order. What we are proposing here is that the implicate order therefore be taken as fundamental. To understand this proposal fully, however, it is necessary to contrast it carefully with what is implied in a mechanistic approach based on the explicate order; for, even in terms of this latter approach, it may of course be admitted that in a certain sense at least, enfoldment and unfoldment can take place in various specific situations (e.g., such as that which happens with the ink droplet). However, this sort of situation is not regarded as having a fundamental kind of significance. All that is primary, independently existent, and universal is thought to be expressible in an explicate order, in terms of elements that are externally related (and these are usually thought to be particles, or fields, or some combination of the two). Whenever enfoldment and unfoldment are found actually to take place, it is therefore assumed that these can ultimately be explained in terms of an underlying explicate order through a deeper mechanical analysis (as, indeed, does happen with the ink-droplet device).

Our proposal to start with the implicate order as basic, then, means that what is primary, independently existent, and universal has to be expressed in terms of the implicate order. So we are suggesting that it is the implicate order that is autonomously active while, as indicated earlier, the explicate order flows out of a law of the implicate order, so that it is secondary, derivative, and appropriate only in certain limited contexts. Or, to put it another way, the relationships constituting the fundamental law are between the enfolded structures that interweave and inter-penetrate each other, throughout the whole of space, rather than between the abstracted and separated forms that are manifest to the senses (and to our instruments).

What, then, is the meaning of the appearance of the apparently independent and self-existent 'manifest world' in the explicate order? The answer to this question is indicated by the root of the word 'manifest', which comes from the Latin 'manus', meaning 'hand'. Essentially, what is manifest is what can be held with the hand – something solid, tangible and visibly stable. The implicate order has its ground in the holomovement which is, as we have seen, vast, rich, and in a state of unending flux of enfoldment and unfoldment, with laws most of which are only vaguely known, and which may even be ultimately unknowable in their totality. Thus it cannot be grasped as something solid, tangible and stable to the senses (or to our instruments). Nevertheless, as has been indicated earlier, the overall law (holonomy) may be assumed to be such that in a certain sub-order, within the whole set of implicate order, there is a totality of forms that have an approximate kind of recurrence, stability and separability. Evidently, these forms are capable of appearing as the relatively solid, tangible, and stable elements that make up our 'manifest world'. The special distinguished sub-order indicated above, which is the basis of the possibility of this manifest world, is then, in effect, what is meant by the explicate order.

We can, for convenience, always picture the explicate order, or imagine it, or represent it to ourselves, as the order present to the senses. The fact that this order *is* actually more or less the one appearing to our senses must, however, be explained. This can be done only when we bring consciousness into our 'universe of discourse' and show that matter in general and consciousness in particular may, at least in a certain sense, have this explicate (manifest) order in common. This question will be explored further when we discuss consciousness in sections 7 and 8.

4 QUANTUM THEORY AS AN INDICATION OF A MULTIDIMENSIONAL IMPLICATE ORDER

Thus far we have been presenting the implicate order as a process of enfoldment and unfoldment taking place in the ordinary three-dimensional space. However, as pointed out in section 2 the quantum theory has a fundamentally new kind of non-local relationship, which may be described as a non-causal connection of elements that are distant from each other, which is brought out in the experiment of Einstein, Podolsky and Rosen.[6] For our purposes, it is not necessary to go into the technical details concerning this non-local relationship. All that is important here is that one finds, through a study of the implications of the quantum theory, that the analysis of a total system into a set of independently existent but interacting particles breaks down in a radically new way. One discovers, instead, both from consideration of the meaning of the mathematical equations and from the results of the actual experiments, that the various particles have to be taken literally as projections of a higher-dimensional reality which cannot be accounted for in terms of any force of interaction between them.[7]

We can obtain a helpful intuitive sense of what is meant by the notion of projection here, through the consideration of the following device. Let us begin with a rectangular tank full of water, with transparent walls (see Figure 7.1). Suppose further that there are two television cameras, A and B, directed at what is going on in the water (e.g., fish swimming around) as seen through the two walls at right angles to each other. Now let the corresponding television images be made visible on screens A and B in another

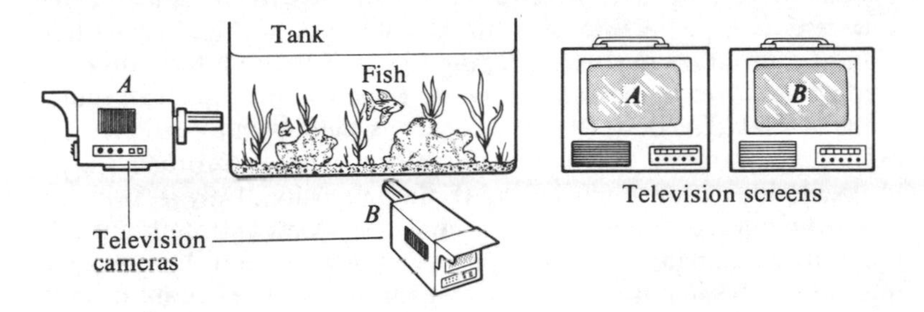

Figure 7.1

room. What we will see there is a certain *relationship* between the images appearing on the two screens. For example, on screen A we may see an image of a fish, and on screen B we will see another such image. At any given moment each image will generally *look* different from the other. Nevertheless the differences will be related, in the sense that when one image is seen to execute certain movements, the other will be seen to execute corresponding movements. Moreover, content that is mainly on one screen will pass into the other, and vice versa (e.g., when a fish initially facing camera A turns through a right angle, the image that was on A is now to be found on B). Thus at all times the image content on the other screen will correlate with and reflect that of the other.

Of course, we know that the two images do not refer to independently existent though interacting actualities (in which, for example, one image could be said to 'cause' related changes in the other). Rather, they refer to a single actuality, which is the common ground of both (and this explains the correlation of images without the assumption that they causally affect each other). This actuality is of higher dimensionality than are the separate images on the screens; or, to put it differently, the images on the screens are two-dimensional *projections* (or facets) of a three-dimensional reality. In some sense this three-dimensional reality holds these two-dimensional projections within it. Yet, since these projections exist only as abstractions, the three-dimensional reality *is* neither of these, but rather it is something else, something of a nature beyond both.

What we are proposing here is that the quantum property of a non-local, non-causal relationship of distant elements may be understood through an extension of the notion described above. That is to say, we may regard each of the 'particles' constituting a system as a projection of a 'higher-dimensional' reality, rather than as a separate particle, existing together with all the others in a common three-dimensional space. For example, in the experiment of Einstein, Podolsky and Rosen, which we have mentioned earlier, each of two atoms that initially combine to form a single molecule are to be regarded as three-dimensional projections of a six-dimensional reality. This may be demonstrated experimentally by causing the molecule to disintegrate and then observing the two atoms after they have separated and are quite distant from each other, so that they do not interact and therefore have no causal connections. What is actually found is that the behaviour of the two atoms is correlated in a way that is rather similar to that of the two television images of the fish, as described earlier. Thus (as is, indeed, further shown by a more careful consideration of the mathematical form of the quantum laws involved here), each electron acts as if it were a projection of a higher-dimensional reality.

Under certain conditions,[8] the two three-dimensional projections corresponding to the two atoms may have a relative independence of behaviour. When these conditions are satisfied it will be a good approximation to treat both atoms as relatively independent but interacting particles, both in the same three-dimensional space. More generally, however, the two atoms will show the typical non-local correlation of behaviour which implies that, more deeply, they are only three-dimensional projections of the kind described above.

A system constituted of N 'particles' is then a $3N$-dimensional reality, of which each 'particle' is a three-dimensional projection. Under the ordinary conditions of our experience, these projections will be close enough to independence so that it will be a good approximation to treat them in the way that we usually do, as a set of separately existing particles all in the same three-dimensional space. Under other conditions this approximation will not be adequate. For example, at low temperatures an aggregate of electrons shows a new property of superconductivity, in which electrical resistance vanishes, so that electric current can flow indefinitely. This is explained by showing that the electrons enter a different kind of state, in which they are no longer relatively independent. Rather, each electron acts as a projection of a single higher-dimensional reality and all these projections share a non-local, non-causal correlation, which is such that they go round obstacles 'co-operatively' without being scattered or diffused, and therefore without resistance. (One could compare this behaviour to a ballet dance, while the usual behaviour of electrons could be compared to that of an agitated crowd of people, moving in a helter-skelter way.)

What follows from all this is that basically the implicate order has to be considered as a process of enfoldment and unfoldment in a higher-dimensional space. Only under certain conditions can this be simplified as a process of enfoldment and unfoldment in three dimensions. Thus far, we have indeed used this sort of simplification, not only with the ink-in-fluid analogy but also with the hologram. Such a treatment, though, is only an approximation, even for the hologram. Indeed, as has already been pointed out earlier in this chapter, the electromagnetic field, which is the ground of the holographic image, obeys the laws of the quantum theory, and when these are properly applied to the field it is found that this, too, is actually a multidimensional reality which can only under certain conditions be simplified as a three-dimensional reality.

Quite generally, then, the implicate order has to be extended into a multidimensional reality. In principle this reality is one unbroken whole, including the entire universe with all its 'fields' and 'particles'. Thus we have to say that the holomovement enfolds and unfolds in a multidimensional order, the dimensionality of which is effectively infinite. However, as we have already seen, relatively independent sub-totalities can generally be abstracted, which may be approximated as autonomous. Thus the principle of relative autonomy of sub-totalities which we introduced earlier as basic to the holomovement is now seen to extend to the multidimensional order of reality.

5 COSMOLOGY AND THE IMPLICATE ORDER

From our consideration of how the general structure of matter can be understood in terms of the implicate order, we now come to certain new notions of cosmology that are implicit in what is being done here.

To bring these out, we first note that when the quantum theory is applied to fields (in the manner discussed in the previous section) it is found that the possible states of energy of this field are discrete (or quantized). Such a state of the field is, in some respects, a wavelike excitation spreading out over a broad region of space. Nevertheless, it also has somehow a discrete quantum of energy (and momentum) proportional to its frequency, so that in other respects it is like a particle[9] (e.g., a photon). However, if one considers the electromagnetic field in empty space, for example, one finds from the quantum theory that each such 'wave-particle' mode of excitation of the field has what is called a 'zero-point' energy, below which it cannot go, even when its energy falls to the minimum that is possible. If one were to add up the energies of all the 'wave-particle' modes of excitation in any region of space, the result would be infinite, because an infinite number of wavelengths is present. However, there is good reason to suppose that one need not keep on adding the energies corresponding to shorter and shorter wavelengths. There may be a certain shortest possible wavelength, so that the total number of modes of excitation, and therefore the energy, would be finite.

Indeed, if one applies the rules of quantum theory to the currently accepted general theory of relativity, one finds that the gravitational field is also constituted of such 'wave-particle' modes, each having a minimum 'zero-point' energy. As a result the gravitational field, and therefore the definition of what is to be meant by distance, cease to be completely defined. As we keep on adding excitations corresponding to shorter and shorter wavelengths to the gravitational field, we come to a certain length at which the measurement of space and time becomes totally undefinable. Beyond this, the whole notion of space and time as we know it would fade out, into something that is at present unspecifiable. So it would be reasonable to suppose, at least provisionally, that this is the shortest wavelength that should be considered as contributing to the 'zero-point' energy of space.

When this length is estimated it turns out to be about 10^{-33} cm. This is much shorter than anything thus far probed in physical experiments (which have got down to about 10^{-17} cm or so). If one computes the amount of energy that would be in one cubic centimetre of space, with this shortest possible wavelength, it turns out to be very far beyond the total energy of all the matter in the known universe.[10]

What is implied by this proposal is that what we call empty space contains an immense background of energy, and that matter as we know it is a small, 'quantized' wavelike excitation on top of this background, rather like a tiny ripple on a vast sea. In current physical theories, one avoids the explicit consideration of this background by calculating only the difference between the energy of empty space and that of space with matter in it. This difference is all that counts in the determination of the general properties of matter as they are presently accessible to observation. However, further developments in physics may make it possible to probe the above-described background in a more direct way. Moreover, even at present, this vast sea of energy may play a key part in the understanding of the cosmos as a whole.

In this connection it may be said that space, which has so much energy, is *full* rather than empty. The two opposing notions of space as empty and space as full have indeed continually alternated with each other in the development of philosophical and physical ideas. Thus, in Ancient Greece, the School of Parmenides and Zeno held that space is a plenum. This view was opposed by Democritus, who was perhaps the first seriously to propose a world view that conceived of space as emptiness (i.e., the void) in which material particles (e.g., atoms) are free to move. Modern science has generally favoured this latter atomistic view, and yet, during the nineteenth century, the former view was also seriously entertained, through the hypothesis of an *ether* that fills all space. Matter, thought of as consisting of special recurrent stable and separable forms in the ether (such as ripples or vortices), would be transmitted through this plenum as if the latter were empty.

A similar notion is used in modern physics. According to the quantum theory, a crystal at absolute zero allows electrons to pass through it without scattering. They go through as if the space were empty. If the temperature is raised, inhomogeneities appear, and these scatter electrons. If one were to use such electrons to observe the crystal (i.e. by focusing them with an electron lens to make an image) what would be visible would be just the inhomogeneities. It would then appear that the inhomogeneities exist independently and that the main body of the crystal was sheer nothingness.

It is being suggested here, then, that what we perceive through the senses as empty space is actually the plenum, which is the ground for the existence of everything, including ourselves. The things that appear to our senses are derivative forms and their true meaning can be seen only when we consider the plenum, in which they are generated and sustained, and into which they must ultimately vanish.

This plenum is, however, no longer to be conceived through the idea of a simple material medium, such as an ether, which would be regarded as existing and moving only in a three-dimensional space. Rather, one is to begin with the holomovement, in which there is the immense 'sea' of energy described earlier. This sea is to be understood in terms of a multidimensional implicate order, along the lines sketched in section 4, while the entire universe of matter as we generally observe it is to be treated as a comparatively small pattern of excitation. This excitation pattern is relatively autonomous and gives rise to approximately recurrent, stable and separable projections into a three-dimensional explicate order of manifestation, which is more or less equivalent to that of space as we commonly experience it.

With all this in mind let us consider the current generally accepted notion that the universe, as we know it, originated in what is almost a single point in space and time from a 'big bang' that happened some ten thousand million years ago. In our approach this 'big bang' is to be regarded as actually just a 'little ripple'. An interesting image is obtained by considering that in the middle of the actual ocean (i.e., on the surface of the Earth) myriads of small waves occasionally come together fortuitously with such phase relationships that they end up in a certain small region of space, suddenly to produce a very high wave which just appears as if from nowhere and out of nothing. Perhaps something like this could happen in the immense ocean of cosmic energy, creating a sudden wave pulse, from which our 'universe' would be born. This pulse would explode outward and break up into smaller ripples that spread yet further outward to constitute our 'expanding universe'. The latter would have its 'space' enfolded within it as a special distinguished explicate and manifest order.[11]

In terms of this proposal it follows that the current attempt to understand our 'universe' as if it were self-existent and independent of the sea of cosmic energy can work at best in some limited way (depending on how far the notion of a relatively independent sub-totality applies to it). For example, the 'black holes' may lead us into an area in which the cosmic background of energy is important. Also, of course, there may be many other such expanding universes.

Moreover, it must be remembered that even this vast sea of cosmic energy takes into account only what happens on a scale larger than the critical length of 10^{-33} cm, to which we have referred earlier. But this length is only a certain kind of limit on the applicability of ordinary notions of space and time. To suppose that there is nothing beyond this limit at all would indeed be quite arbitrary. Rather, it is very probable that beyond it lies a further domain, or set of domains, of the nature of which we have as yet little or no idea.

What we have seen thus far is a progression from explicate order to simple three-dimensional implicate order, then to a multi-dimensional implicate order, then to an extension of this to the immense 'sea' in what is sensed as empty space. The next stage may well lead to yet further enrichment and extension of the notion of implicate order, beyond the critical limit of 10^{-33} cm mentioned above; or it may lead to some basically new notions which could not be comprehended even within the possible further developments of the implicate order. Nevertheless, whatever may be possible in this regard, it is clear that we may assume that the principle of relative autonomy of sub-totalities continues to be valid. Any sub-totality, including those which we have thus far considered, may up to a point be studied in its own right. Thus, without assuming that we have already arrived even at an outline of absolute and final truth, we may at least for a time put aside the need to consider what may be beyond the immense energies of empty space, and go on to bring out the further implications of the sub-totality of order that has revealed itself thus far.

6 THE IMPLICATE ORDER, LIFE AND THE FORCE OF OVERALL NECESSITY

In this section we shall bring out the meaning of the implicate order by first showing how it makes possible the comprehension of both inanimate matter and life on the basis of a single ground, common to both, and then we shall go on to propose a certain more general form for the laws of the implicate order.

Let us begin by considering the growth of a living plant. This growth starts from a seed, but the seed contributes little or nothing to the actual material substance of the plant or to the energy needed to make it grow. This latter comes almost entirely from the soil, the water, the air and the sunlight. According to modern theories the seed contains *information*, in the form of DNA, and this information somehow 'directs' the environment to form a corresponding plant.

In terms of the implicate order, we may say that even inanimate matter maintains itself in a continual process similar to the growth of plants. Thus, recalling the ink-in-fluid model of the electron, we see that such a 'particle' is to be understood as a recurrent stable order of unfoldment in which a certain form undergoing regular changes manifests again and again, but so rapidly that it appears to be in continuous existence. We may compare this to a forest, constituted of trees that are continually dying and being replaced by new ones. If it is considered on a long time-scale, this forest may be regarded likewise as a continuously existent but slowly-changing entity. So when understood through the implicate order, inanimate matter and living beings are seen to be, in certain key respects, basically similar to their modes of existence.

When inanimate matter is left to itself the above-described process of enfoldment and unfoldment just reproduces a similar form of inanimate matter, but when this is further 'informed' by the seed, it begins to produce a living plant instead. Ultimately, this latter gives rise to a new seed, which allows the process to continue after the death of this plant.

As the plant is formed, maintained and dissolved by the exchange of matter and energy with its environment, at which point can we say that there is a sharp distinction between what is alive and what is not? Clearly, a molecule of carbon dioxide that crosses a cell boundary into a leaf does not suddenly 'come alive' nor does a molecule of oxygen suddenly 'die' when it is released to the atmosphere. Rather, life itself has to be regarded as belonging in some sense to a totality, including plant and environment.

It may indeed be said that life is enfolded in the totality and that, even when it is not manifest, it is somehow 'implicit' in what we generally call a situation in which there is no life. We can illustrate this by considering the ensemble of all the atoms that are now in the environment but that are eventually going to constitute a plant that will grow from a certain seed. This ensemble is evidently, in certain key ways, similar to that considered in section 3, of ink particles forming a droplet. In both cases the elements of the ensemble are bound together to contribute to a common end (in one case an ink droplet and in the other case a living plant).

The above does not mean, however, that life can be reduced

completely to nothing more than that which comes out of the activity of a basis governed by the laws of inanimate matter alone (though we do not deny that *certain* features of life may be understood in this way). Rather, we are proposing that as the notion of the holomovement was enriched by going from three-dimensional to multidimensional implicate order and then to the vast 'sea' of energy in 'empty' space, so we may now enrich this notion further by saying that in its totality the holomovement includes the principle of life as well. Inanimate matter is then to be regarded as a relatively autonomous sub-totality in which, at least as far as we now know, life does not significantly manifest. That is to say, inanimate matter is a secondary, derivative, and particular abstraction from the holomovement (as would also be the notion of a 'life force' entirely independent of matter). Indeed, the holomovement which is 'life implicit' is the ground both of 'life explicit' and of 'inanimate matter', and this ground is what is primary, self-existent and universal. Thus we do not fragment life and inanimate matter, nor do we try to reduce the former completely to nothing but an outcome of the latter.

Let us now put the above approach in a more general way. What is basic to the law of the holomovement is, as we have seen, the possibility of abstraction of a set of relatively autonomous sub-totalities. We can now add that the laws of each such abstracted sub-totality quite generally operate under certain conditions and limitations defined only in a corresponding total situation (or set of similar situations). This operation will in general have these three key features:

1. A set of implicate orders.
2. A special distinguished case of the above set, which constitutes an explicate order of manifestation.
3. A general relationship (or law) expressing a force of necessity which binds together a certain set of the elements of the implicate order in such a way that they contribute to a common explicate end (different from that to which another set of inter-penetrating and intermingling elements will contribute).

The origin of this force of necessity cannot be understood solely in terms of the explicate and implicate orders belonging to the type of situation in question. Rather, at this level, such necessity has simply to be accepted as inherent in the overall situation under discussion. An understanding of its origin would take us to a deeper, more comprehensive and more inward level of relative autonomy which, however, would also have its implicate and explicate orders and a correspondingly deeper and more inward force of necessity that would bring about their transformation into each other.[12]

In short, we are proposing that this *form* of the law of a relatively autonomous sub-totality, which is a consistent generalization of all the forms that we have studied thus far, is to be considered as universal; and that in our subsequent work we shall explore the implicates of such a notion, at least tentatively and provisionally.

7 CONSCIOUSNESS AND THE IMPLICATE ORDER

At this point it may be said that at least some outlines of our notions of cosmology and of the general nature of reality have been sketched (though, of course, to 'fill in' this sketch with adequate detail would require a great deal of further work much of which still remains to be done). Let us now consider how consciousness may be understood in relation to these notions.

We begin by proposing that in some sense, consciousness (which we take to include thought, feeling, desire, will, etc.) is to be comprehended in terms of the implicate order, along with reality as a whole. That is to say, we are suggesting that the implicate order applies both to matter (living and non-living) and to consciousness, and that it can therefore make possible an understanding of the general relationship of these two, from which we may be able to come to some notion of a common ground of both (rather as was also suggested in the previous section in our discussion of the relationship of inanimate matter and life).

To obtain an understanding of the relationship of matter and consciousness has, however, thus far proved to be extremely difficult, and this difficulty has its root in the very great difference in their basic qualities, as they present themselves in our experience. This difference has been expressed with particularly great clarity by Descartes, who described matter as 'extended substance' and consciousness as 'thinking substance'. Evidently, by 'extended substance' Descartes meant something made up of distinct forms existing in space, in an order of extension and separation basically similar to the one that we have been calling explicate. By using the term 'thinking substance' in such sharp contrast to 'extended substance' he was clearly implying that the various distinct forms appearing in thought do not have their existence in such an order of extension and separation (i.e., some kind of space), but rather in a different order, in which extension and separations have no fundamental significance. The implicate order has just this latter quality, so in a certain sense Descartes was perhaps anticipating that consciousness has to be understood in terms of an order that is closer to the implicate than it is to the explicate.

However, when we start, as Descartes did with extension and separation in space as primary for matter, then we can see nothing in this notion that can serve as a basis for a relationship between matter and consciousness, whose orders are so different. Descartes clearly understood this difficulty and indeed proposed to resolve it by means of the idea that such a relationship is made possible by God, who being outside of and beyond matter and consciouness (both of which He has indeed created) is able to give the latter 'clear and distinct notions' that are currently applicable to the former. Since then, the idea that God takes care of this requirement has generally been abandoned, but it has not commonly been noticed that thereby the possibility of comprehending the relationship between matter and consciousness has collapsed.

In this chapter, we have, however, shown in some detail that matter as a whole can be understood in terms of the notion that the implicate order is the immediate and primary actuality (while the explicate order can be derived as a particular, distinguished case of the implicate order). The question that arises here, then, is that of whether or not (as was in a certain sense anticipated by Descartes) the actual 'substance' of consciousness can be understood in terms of the notion that the implicate order is also its primary and immediate actuality. If matter and consciousness could in this way be understood together, in terms of the same general notion of order, the way would be opened to comprehending their relationship on the basis of some common ground.[13] Thus we could come to the germ of a new notion of unbroken wholeness, in which consciousness is no longer to be fundamentally separated from matter.

Let us now consider what justification there is for the notion that matter and consciousness have the implicate order in common. First, we note that matter in general is, in the first instance, the object of our consciousness. However, as we have seen throughout this chapter, various energies such as light, sound, etc., are continually enfolding information in principle concerning the entire universe of matter into each region of space. Through this process, such information may of course enter our sense organs, to go on through the nervous system to the brain. More deeply, all the matter in our bodies, from the very first, enfolds the universe in some way. Is this enfolded structure, both of information and of matter (e.g., in the brain and nervous system), that which primarily enters consciousness?

Let us first consider the question of whether information is actually enfolded in the brain cells. Some light on this question is afforded by certain work on brain structure, notably that of Pribram.[14] Pribram has given evidence backing up his suggestion that memories are generally recorded all over the brain, in such a way that information concerning a given object or quality is not stored in a particular cell or localized part of the brain but rather that all the information is enfolded over the whole. This storage resembles a hologram in its function, but its actual structure is much more complex. We can then suggest that when the 'holographic' record in the brain is suitably activated, the response is to create a pattern of nervous energy constituting a partial experience similar to that which produced the 'hologram' in the first place. But it is also different in that it is less detailed, in that memories from many different times may merge together, and in that memories may be connected by association and by logical thought to give a certain further order to the whole pattern. In addition, if sensory data is also being attended to at the same time, the whole of this response from memory will, in general, fuse with the nervous excitation coming from the senses to give rise to an overall experience in which memory, logic, and sensory activity combine into a single unanalysable whole.

Of course, consciousness is more than what has been described above. It also involves awareness, attention, perception, acts of understanding, and perhaps yet more. We have suggested in the first chapter that these must go beyond a mechanistic response (such as that which the holographic model of brain function would by itself imply). So in studying them we may be coming closer to the essence of actual conscious experience than is possible merely by discussing patterns of excitation of the sensory nerves and how they may be recorded in memory.

It is difficult to say much about faculties as subtle as these. However, by reflecting on and giving careful attention to what happens in certain experiences, one can obtain valuable clues. Consider, for example, what takes place when one is listening to music. At a given moment a certain note is being played but a number of the previous notes are still 'reverberating' in consciousness. Close attention will show that it is the simultaneous presence and activity of all these reverberations that is responsible for the direct and immediately felt sense of movement, flow and continuity. To hear a set of notes so far apart in time that there is no such reverberation will destroy altogether the sense of a whole unbroken, living movement that gives meaning and force to what is heard.

It is clear from the above that one does not experience the actuality of this whole movement by 'holding on' to the past, with the aid of a memory of the sequence of notes, and comparing this past with the present. Rather, as one can discover by further attention, the 'reverberations' that make such an experience possible are not memories but are rather *active transformations* of what came earlier, in which are to be found not only a generally diffused sense of the original sounds, with an intensity that falls off, according to the time elapsed since they were picked up by the ear, but also various emotional responses, bodily sensations, incipient muscular movements, and the evocation of a wide range of yet further meanings, often of great subtlety. One can thus obtain a direct sense of how a sequence of notes is enfolding into many levels of consciousness, and of how at any given moment, the transformations flowing out of many such enfolded notes interpenetrate and intermingle to give rise to an immediate and primary feeling of movement.

This activity in consciousness evidently constitutes a striking parallel to the activity that we have proposed for the implicate order in general. Thus in section 3, we have given a model of an electron in which, at any instant, there is a co-present set of differently transformed ensembles which inter-penetrate and intermingle in their various degrees of enfoldment. In such enfoldment, there is a radical change, not only of form but also of structure, in the entire set of ensembles (which change we have, in chapter 6, called a metamorphosis); and yet, a certain totality of order in the ensembles remains invariant, in the sense that in all these changes a subtle but fundamental similarity of order is preserved.[15]

In the music, there is, as we have seen, a basically similar transformation (of notes) in which a certain order can also be seen to be preserved. The key difference in these two cases is that for our model of the electron an enfolded order is grasped *in thought*, as the presence together of many different but interrelated degrees of transformations of ensembles, while for the music, it is *sensed immediately* as the presence together of many different but interrelated degrees of transformations of tones and sounds. In the latter, there is a feeling of both tension and harmony between the various co-present transformations, and this feeling is indeed what is primary in the apprehension of the music in its undivided state of flowing movement.

In listening to music, *one is therefore directly perceiving an implicate order*. Evidently this order is *active* in the sense that it continually flows into emotional, physical, and other responses, that are inseparable from the transformations out of which it is essentially constituted.

A similar notion can be seen to be applicable for vision. To bring this out, consider the sense of motion that arises when one is watching the cinema screen. What is actually happening is that a series of images, each slightly different, is being flashed on the screen. If the images are separated by long intervals of time, one does not get a feeling of continuous motion, but rather, one sees a series of disconnected images perhaps accompanied by a sense of jerkiness. If, however, the images are close enough together (say a hundredth of a second) one has a direct and immediate experience, as if from a continuously moving and flowing reality, undivided and without a break.

This point can be brought out even more clearly by considering a well-known illusion of movement, produced with the aid of a stroboscopic device, illustrated in figure 7.2.

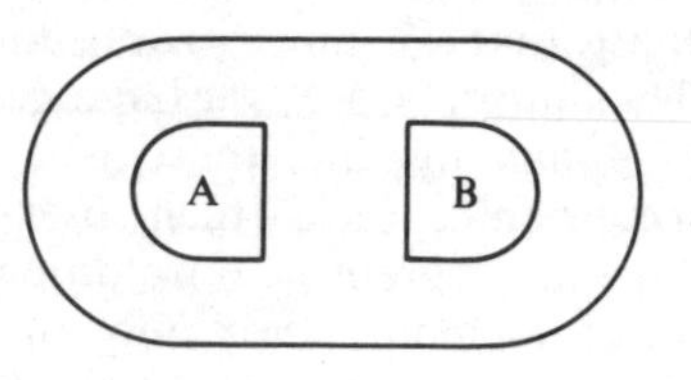

Figure 7.2

Two discs, A and B, enclosed in a bulb, can be caused to give off light by means of electrical excitation. The light is made to flash on and off so rapidly that it appears to be continuous, but in each flash it is arranged that B will come on slightly later than A. What one actually feels is a sense of 'flowing movement' between A and B, but that paradoxically nothing is flowing out of B (contrary to what would be expected if there had been a real process of flow). This means that a sense of flowing movement is experienced when, on the retina of the eye, there are two images in neighbouring positions one of which comes on slightly later than the other.

(Closely related to this is the fact that a blurred photograph of a speeding car, containing a sequence of overlaid images in slightly different positions, conveys to us a much more immediate and vivid sense of movement than does a sharp picture, taken with a high-speed camera.)

It seems evident that the sense of unbroken movement described above is basically similar to that arising from a sequence of musical notes. The main difference between music and visual images, in this regard, is that the latter may arrive so close together in time that they cannot be resolved in consciousness. Nevertheless, it is clear that visual images must also undergo active transformation as they 'enfold' into the brain and nervous system (e.g., they give rise to emotional, physical and other more subtle responses of which one may be only dimly conscious as well as to 'after images' that are in certain ways similar to the reverberations in musical notes). Even though the time difference of two such images may be small, the examples cited above make it clear that a sense of movement is experienced through the intermingling and inter-penetration of the co-present transformations to which these images must give rise, as they penetrate the brain and nervous system.

All of this suggests that quite generally (and not merely for the special case of listening to music), there is a basic similarity between the order of our immediate experience of movement and the implicate order as expressed in terms of our thought. We have in this way been brought to the possibility of a coherent mode of understanding the immediate experience of motion in terms of our thought (in effect thus resolving Zeno's paradox concerning motion).

To see how this comes about, consider how motion is usually thought of, in terms of a series of points along a line. Let us suppose that at a certain time t_1, a particle is at a position x_1, while at a later time t_2, it is at another position x_2. We then say that this particle is moving and that its velocity is

$$v = \frac{x_2 - x_1}{t_2 - t_1} \ .$$

Of course, this way of thinking does not in any way reflect or convey the immediate sense of motion that we may have at a given moment, for example, with a sequence of musical notes reverberating in consciousness (or in the visual perception of a speeding car). Rather, it is only an abstract symbolization of movement, having a relation to the actuality of motion, similar to that between a musical score and the actual experience of the music itself.

If, as is commonly done, we take the above abstract symbolization as a faithful representation of the actuality of movement we become entangled in a series of confused and basically insoluble problems. These all have to do with the image in which we represent time, as if it were a series of points along a line that are somehow all present together, either to our conceptual gaze or perhaps to that of God. Our actual experience is, however, that when a given moment, say t_2, is present and actual, an earlier moment, such as t_1 is past. That is to say, it is *gone*, non-existent, never to return. So if we say that the velocity of a particular *now* (at t_2) is $(x_2 - x_1)/(t_2 - t_1)$ we are trying to relate *what is* (i.e., x_2 and t_2) to *what is not* (i.e., x_1 and t_1). We can of course do this *abstractly and symbolically* (and is, indeed, the common practice in science and mathematics), but the further fact, not comprehended in this abstract symbolism, is that the velocity *now* is active *now* (e.g., it determines how a particle will act from now on, in itself, and in relation to other particles). How are we to understand the *present* activity of a position (x_1) that is now non-existent and gone for ever?

It is commonly thought that this problem is resolved by the differential calculus. What is done here is to let the time interval, $\Delta t = t_2 - t_1$ become vanishingly small, along with $\Delta x = x_2 - x_1$. The velocity *now* is defined as the limit of the ratio $\Delta x/\Delta t$ as Δt approaches zero. It is then implied that the problem described above no longer arises, because x_2 and x_1 are in effect taken at the same time. They may thus be present together and related in an activity that depends on both.

A little reflection shows, however, that this procedure is still as abstract and symbolic as was the original one in which the time interval was taken as finite. Thus one has no immediate experience of a time interval of zero length, nor can one see in terms of reflective thought what this could mean.

Even as an abstract formalism, this approach is not fully consistent in a logical sense, nor does it have a universal range of applicability. Indeed, it applies only within the area of *continuous* movements and then only as a technical algorithm what happens to be correct for this sort of movement. As we have seen, however, according to the quantum theory, movement is *not* fundamentally continuous. So even as an algorithm its current field of application is limited to theories expressed in terms of classical concepts (i.e., in the explicate order) in which it provides a good approximation for the purpose of calculating the movements of material objects.

When we think of movement in terms of the implicate order,[16] however, these problems do not arise. In this order, movement is comprehended in terms of a series of inter-penetrating and intermingling elements in different degrees of enfoldment *all present together*. The activity of this movement then presents no difficulty, because it is an outcome of this whole enfolded order, and is determined by relationships of co-present elements, rather than by the relationships of elements that exist to others that no longer exist.

We see, then, that through thinking in terms of the implicate order, we come to a notion of movement that is logically coherent and that properly represents our immediate experience of movement. Thus the sharp break between abstract logical thought and concrete immediate experience, that has pervaded our culture for so long, need no longer be maintained. Rather, the possibility is created for an unbroken flowing movement from immediate experience to logical thought and back, and thus for an ending to this kind of fragmentation.

Moreover we are now able to understand in a new and more consistent way our proposed notion concerning the general nature of reality, that *what is* is movement. Actually, what tends to make it difficult for us to work in terms of this notion is that we usually think of movement in the traditional way as an active relationship of what is to what is not. Our traditional notion concerning the general nature of reality would then amount to saying that *what is* is an active relationship of what is to what is not. To say this is, at the very least, confused. In terms of the implicate order, however, movement is a relationship of certain phases of *what is* to other phases of *what is*, that are in different stages of enfoldment. This notion implies that the essence of reality as a whole is the above relationship among the various phases in different stages of enfoldment (rather than, for example, a relationship between various particles and fields that are all explicate and manifest).

Of course, actual movement involves more than the mere immediate intuitive sense of unbroken flow, which is our mode of directly experiencing the implicate order. The presence of such a sense of flow generally implies further that, in the next moment, the state of affairs will actually change – i.e., it will be different. How are we to understand this fact of experience in terms of the implicate order?

A valuable clue is provided by reflecting on and giving careful attention to what happens when, in our thinking, we say that one set of ideas *implies* an entirely different set. Of course, the word 'imply' has the same root as the word 'implicate' and thus also involves the notion of enfoldment. Indeed, by saying that something is *implicit* we generally mean more than merely to say that this thing is an inference following from something else through the rules of logic. Rather, we usually mean that from many different ideas and notions (of some of which we are explicity conscious) a new notion emerges that somehow brings all these together in a concrete and undivided whole.

We see, then, that each moment of consciousness has a certain *explicit* content, which is a foreground, and an *implicit* content, which is a corresponding background. We now propose that not only is immediate experience best understood in terms of the implicate order, but that thought also is basically to be comprehended in this order. Here we mean not just the *content* of thought for which we have already begun to use the implicate order. Rather, we also mean that the actual *structure, function* and *activity* of thought is in the implicate order. The distinction between implicit and explicit in thought is thus being taken here to be essentially equivalent to the distinction between implicate and explicate in matter in general.

To help clarify what this means, let us recall briefly the basic form of the law of a sub-totality (discussed in sections 3 and 6), i.e., that the enfolded elements of a characteristic ensemble (e.g., of ink particles or of atoms) that are going to constitute the next stage of enfoldment are bound by a force of overall necessity, which brings them together, to contribute to a common end that emerges in the next phase of the process under discussion. Similarly, we propose that the ensemble of elements enfolded in the brain and nervous system that are going to constitute the next stage of development of a line of thought are likewise bound through a force of overall necessity, which brings them together to contribute to the common notion that emerges in the next moment of consciousness.

In this study, we have been using the idea that consciousness can be described in terms of a series of moments. Attention shows that a given moment cannot be fixed exactly in relation to time (e.g., by the clock) but rather, that it covers some vaguely defined and somewhat variable extended period of duration. As pointed out earlier, each moment is experienced directly in the implicate order. We have further seen that through the force of necessity in the overall situation, one moment gives rise to the next, in which content that was previously implicate is now explicate while the previous explicate content has become implicate (e.g., as happened in the analogy of the ink droplets).

The continuation of the above process gives an account of how *change* takes place from one moment to another. In principle, the change in any moment may be a fundamental and radical transformation. However, experience shows that in thought (as in matter in general) there is usually a great deal of recurrence and stability leading to the possibility of relatively independent sub-totalities.

In any such sub-totality, there is the possibility of the continuation of a certain line of thought, that enfolds in a fairly regularly changing way. Evidently, the precise character of such a sequence of thoughts, as it enfolds from one moment to the next, will generally depend on the content of the implicate order in earlier moments. For example, a moment containing a sense of movement tends quite generally to be followed by a change in the next moment, which is greater the stronger the sense of movement that was originally present (so that, as in the case of the stroboscopic device discussed earlier, when this does not happen we feel that something surprising or paradoxical is taking place).

As in our discussion of matter in general, it is now necessary to go into the question of how in consciousness the explicate order is what is manifest. As observation and attention show (keeping in mind that the word 'manifest' means that which is recurrent, stable and separable) the manifest content of consciousness is based essentially on memory, which is what allows such content to be held in a fairly constant form. Of course, to make possible such constancy it is also necessary that this content be organized, not only through relatively fixed associations but also with the aid of the rules of logic, and of our basic categories of space, time, causality, universality, etc. In this way an overall system of concepts and mental images may be developed, which is a more or less faithful representation of the 'manifest world'.

The process of thought is not, however, merely a *representation* of the manifest world; rather, it makes an important *contribution* to how we experience this world, for, as we have already pointed out earlier, this experience is a fusion of sensory information with the 'replay' of some of the content of memory (which latter contains thought built into its very form and order). In such experience, there will be a strong background of recurrent stable, and separable features, against which the transitory and changing aspects of the unbroken flow of experience will be seen as fleeting impressions that tend to be arranged and ordered mainly in terms of the vast totality of the relatively static and fragmented content of recordings from the past.

One can, in fact, adduce a considerable amount of scientific evidence showing how much of our conscious experience is a construction based on memory organized through thought, in the general way described above.[17] To go into this subject in detail would, however, carry us too far afield. It may nevertheless be useful here to mention that Piaget[18] has made it clear that a conscioùness of what to us is the familiar order of space, time, causality, etc. (which is essentially what we have been calling the explicate order) operates only to a small extent in the earliest phases of life of the human individual. Rather, as he shows from careful observations, for the most part infants *learn* this content first in the area of sensori-motor experience, and later as they grow older they connect such experience with its expression in language and logic. On the other hand, there seems to be an immediate awareness of movement from the very earliest. Recalling that movement is sensed primarily in the implicate order, we see that Piaget's work supports the notion that the experiencing of the implicate order is fundamentally much more immediate and direct than is that of the explicate order, which, as we have pointed out above, requires a complex construction that has to be learned.

One reason why we do not generally notice the primacy of the implicate order is that we have become so habituated to the explicate order, and have emphasized it so much in our thought and language, that we tend strongly to feel that our primary experience is of that which is explicate and manifest. However, another reason, perhaps more important, is that the activation of memory recordings whose content is mainly that which is recurrent, stable, and separable, must evidently focus our attention very strongly on what is static and fragmented.

This then contributes to the formation of an experience in which these static and fragmented features are often so intense that the more transitory and subtle features of the unbroken flow (e.g., the 'transformations' of musical notes) generally tend to pale into such seeming insignificance that one is, at best, only dimly conscious of them. Thus, an illusion may arise in which the manifest static and fragmented content of consciousness is experienced as the very basis of reality and from this illusion one may apparently obtain a proof of the correctness of that mode of thought in which this content is taken to be fundamental.[19]

8 MATTER, CONSCIOUSNESS AND THEIR COMMON GROUND

At the beginning of the previous section we suggested that matter and consciousness can both be understood in terms of the implicate order. We shall now show how the notions of implicate order that we have developed in connection with consciousness may be related to those concerning matter, to make possible an understanding of how both may have a common ground.

We begin by noting that (as pointed out in chapters 1 and 5) current relativistic theories in physics describe the whole of reality in terms of a process whose ultimate element is a point event, i.e., something happening in a relatively small region of space and time. We propose instead that the basic element be a *moment* which, like the moment of consciousness, cannot be precisely related to measurements of space and time, but rather covers a somewhat vaguely defined region which is extended in space and has duration in time. The extent and duration of a moment may vary from something very small to something very large, according to the context under discussion (even a particular century may be a 'moment' in the history of mankind). As with consciousness, each moment has a certain explicate order, and in addition it enfolds all the others, though in its own way. So the relationship of each moment in the whole to all the others is implied by its total content: the way in which it 'holds' all the others enfolded within it.

In certain ways this notion is similar to Leibniz's idea of monads, each of which 'mirrors' the whole in its own way, some in great detail and others rather vaguely. The difference is that Leibniz's monads had a permanent existence, whereas our basic elements are only moments and are thus not permanent. Whitehead's idea of 'actual occasions' is closer to the one proposed here, the main difference being that we use the implicate order to express the qualities and relationships of our moments, whereas Whitehead does this in a rather different way.

We now recall that the laws of the implicate order are such that there is a relatively independent, recurrent, stable sub-totality which constitutes the explicate order, and which, of course, is basically the order that we commonly contact in common experience (extended in certain ways by our scientific instruments). This order has room in it for something like memory, in the sense that previous moments generally leave a trace (usually enfolded) that continues in later moments, though this trace may change and transform almost without limit. From this trace (e.g., in the rocks) it is in principle possible for us to unfold an image of past moments, similar in certain ways, to what actually happened; and by taking advantage of such traces, we design instruments such as photographic cameras, tape recorders, and computer memories, which are able to register actual moments in such a way that much more of the content of what has happened can be made directly and immediately accessible to us, than is generally possible from natural traces alone.

One may indeed say that our memory is a special case of the process described above, for all that is recorded is held enfolded within the brain cells and these are part of matter in general. The recurrence and stability of our own memory as a relatively independent sub-totality is thus brought about as part of the very same process that sustains the recurrence and stability in the manifest order of matter in general.

It follows, then, that the explicate and manifest order of consciousness is not ultimately distinct from that of matter in general. Fundamentally these are essentially different aspects of the one overall order. This explains a basic fact that we have pointed out earlier – that the explicate order of matter in general is also in essence the sensuous explicate order that is presented in consciousness in ordinary experience.

Not only in this respect but, as we have seen, also in a wide range of other important respects, consciousness and matter in general are basically the same order (i.e., the implicate order as a whole). As we have indicated earlier this order is what makes a relationship between the two possible; but more specifically, what are we to say about the nature of this relationship?

We may begin by considering the individual human being as a relatively independent sub-totality, with a sufficient recurrence and stability of his total process (e.g., physical, chemical, neurological, mental, etc.) to enable him to subsist over a certain period of time. In this process we know it to be a fact that the physical state can affect the content of consciousness in many ways. (The simplest case is that we can become conscious of neural excitations as sensations). Vice versa, we know that the content of consciousness can affect the physical state (e.g., from a conscious intention nerves may be excited, muscles may move, the heart-beat change, along with alterations of glandular activity, blood chemistry, etc.).

This connection of the mind and body has commonly been called psychosomatic (from the Greek 'psyche', meaning 'mind' and 'soma', meaning 'body'). This word is generally used, however, in such a way as to imply that mind and body are separately existent but connected by some sort of interaction. Such a meaning is not compatible with the implicate order. In the implicate order we have to say that mind enfolds matter in general and therefore the body in particular. Similarly, the body enfolds not only the mind but also in some sense the entire material universe. (In the manner explained earlier in this section, both through the senses and through the fact that the constituent atoms of the body are actually structures that are enfolded in principle throughout all space.)

This kind of relationship has in fact already been encountered in section 4, where we introduced the notion of a higher-dimensional reality, which *projects* into lower-dimensional elements that have not only a non-local and non-causal relationship but also just the sort of mutual enfoldment that we have suggested for mind and body. So we are led to propose further that the more comprehensive, deeper, and more inward actuality is neither mind nor body but rather a yet higher-dimensional actuality, which is their common ground and which is of a nature beyond both. Each of these is then only a relatively independent sub-totality and it is implied that this relative independence derives from the higher-dimensional ground in which mind and body are ultimately one (rather as we find that the relative independence of the manifest order derives from the ground of the implicate order).

In this higher-dimensional ground the implicate order prevails. Thus, within this ground, *what is* is movement which is represented in thought as the co-presence of many phases of the implicate order. As happens with the simpler forms of the implicate order considered earlier, the state of movement at one moment unfolds through a more inward force of necessity inherent in this overall state of affairs, to give rise to a new state of affairs in the next moment. The projections of the higher-dimensional ground, as mind and body, will be in the later moment both be different from what they were in the earlier moment, though these differences will of course be related. So we do not say that mind and body causally affect each other, but rather than the movements of both are the outcome of related projections of a common higher-dimensional ground.

Of course, even this ground of mind and body is limited. At the very least we have evidently to include matter beyond the body if we are to give an adequate account of what actually happens and this must eventually include other people, going on to society and to mankind as a whole. In doing this, however, we will have to be careful not to slip back into regarding the various elements of any given total situation as having anything more than relative independence. In a deeper and generally more suitable way of thinking, each of these elements is a projection, in a sub-totality of yet higher 'dimension'. So it will be ultimately misleading and indeed wrong to suppose, for example, that each human being is an independent actuality who interacts with other human beings and with nature. Rather, all these are projections of a single totality. As a human being takes part in the process of this totality, he is fundamentally changed in the very activity in which his aim is to change that reality which is the content of his consciousness. To fail to take this into account must inevitably lead one to serious and sustained confusion in all that one does.

From the side of mind we can also see that it is necessary to go on to a more inclusive ground. Thus, as we have seen, the easily accessible explicit content of consciousness is included within a much greater implicit (or implicate) background. This in turn evidently has to be contained in a yet greater background which may include not only neuro-physiological processes at levels of which we are not generally conscious but also a yet greater background of unknown (and indeed ultimately unknowable) depths of inwardness that may be analogous to the 'sea' of energy that fills the sensibly perceived 'empty' space.[20]

Whatever may be the nature of these inward depths of consciousness, they are the very ground, both of the explicit content and of that content which is usually called implicit. Although this ground may not appear in ordinary consciousness, it may nevertheless be present in a certain way. Just as the vast 'sea' of energy in space is present to our perception as a *sense* of emptiness or nothingness so the vast 'unconscious' background of explicit consciousness with all its implications is present in a similar way. That is to say, it may be *sensed* as an emptiness, a nothingness, within which the usual content of consciousness is only a vanishingly small set of facets.

Let us now consider briefly what may be said about time in this total order of matter and consciousness.

First, it is well known that as directly sensed and experienced in consciousness, time is highly variable and relative to conditions (e.g., a given period may be felt to be short or long by different people, or even by the same person, according to the interests of the different people concerned). On the other hand it seems in common experience that physical time is absolute and does not depend on conditions. However, one of the most important implications of the theory of relativity is that physical time is in fact relative, in the sense that it may vary according to the speed of the observer. (This variation is, however, significant only as we approach the speed of light and is quite negligible in the domain of ordinary experience.) What is crucial in the present context is that, according to the theory of relativity, a sharp distinction between space and time can not be maintained (except as an approximation, valid at velocities small compared with that of light). Thus, since the quantum theory implies that elements that are separated in space are generally non-causally and non-locally related projections of a higher-dimensional reality, it follows that moments separated in time are also such projections of this reality.

Evidently, this leads to a fundamentally new notion of the meaning of time. Both in common experience and in physics, time has generally been considered to be a primary, independent and universally applicable order, perhaps the most fundamental one known to us. Now, we have been led to propose that it is secondary and that, like space (see section 5), it is to be derived from a higher-dimensional ground, as a particular order. Indeed, one can further say that many such particular interrelated time orders can be derived for different sets of sequences of moments, corresponding to material systems that travel at different speeds. However, these are all dependent on a multidimensional reality that cannot be comprehended fully in terms of any time order, or set of such orders.

Similarly, we are led to propose that this multidimensional reality may project into many orders of sequences of moments in consciousness. Not only do we have in mind here the relativity of psychological time discussed above, but also much more subtle implications. Thus, for example, people who know each other well may separate for a long time (as measured by the sequence of moments registered by a clock) and yet they are often able to 'take up from where they left off' as if no time had passed. What we are proposing here is that sequences of moments that 'skip' intervening spaces are just as allowable forms of time as those which seem continuous.[21]

The fundamental law, then, is that of the immense multidimensional ground; and the projections from this ground determine whatever time orders there may be. Of course, this law may be such that in certain limiting cases the order of moments corresponds approximately to what would be determined by a simple causal law. Or, in a different limiting case, the order would be a complex one of a high degree which would, as indicated in chapter 5, approximate what is usually called a random order. These two alternatives cover what happens for the most part in the domain of ordinary experience as well as in that of classical physics. Nevertheless in the quantum domain as well as in connection with consciousness and probably with the understanding of the deeper more inward essence of life, such approximations will prove to be inadequate. One must then go on to a consideration of time as a projection of multidimensional reality into a sequence of moments.

Such a projection can be described as creative, rather than mechanical, for by creativity one means just the inception of new content, which unfolds into a sequence of moments that is not completely derivable from what came earlier in this sequence or set of such sequences. What we are saying is, then, that movement is basically such a creative inception of new content as projected from the multidimensional ground and that in contrast what is mechanical is a relatively autonomous sub-totality that can be abstracted from that which is basically a creative movement of unfoldment.

How, then, are we to consider the evolution of life as this is generally formulated in biology? First, it has to be pointed out that the very word 'evolution' (whose literal meaning is 'unrolling') is too mechanistic in its connotation to serve properly in this context. Rather, as we have already pointed out above, we should say that various successive living forms unfold creatively and in the sense that later members are not completely derivable from what came earlier, through a process in which effect arises out of cause (though in some approximation such a causal process may explain certain limited aspects of the sequence). The law of this unfoldment cannot be properly understood without considering the immense multidimensional reality of which it is a projection (except in the rough approximation in which the implications of the quantum theory and of what is beyond this theory may be neglected).

Our overall approach has thus brought together questions of the nature of the cosmos, of matter in general, of life, and of consciousness. All of these have been considered to be projections of a common ground. This we may call the ground of all that is, at least in so far as this may be sensed and known by us, in our present phase of unfoldment of consciouness. Although we have no detailed perception or knowledge of this ground it is still in a certain sense enfolded in our consciouness, in the ways in which we have outlined, as well as perhaps in other ways that are yet to be discovered.

Is this ground the absolute end of everything? In our proposed views concerning the general nature of 'the totality of all that is' we regard even this ground as a mere stage, in the sense that there could in principle be an infinity of further development beyond it. At any particular moment in this development each such set of views that may arise will constitute at most a *proposal*. It is not to be taken as an *assumption* about what the final truth is supposed to be, and still less as a *conclusion* concerning the nature of such truth. Rather, this proposal becomes itself an *active factor* in the totality of existence which includes ourselves as well as the objects of our thoughts and experimental investigations. Any further proposals on this process will, like those already made, have to be *viable*. That is to say, one will require of them a general self-consistency as well as consistency in what flows from them in life as a whole. Through the force of an even deeper more inward necessity in this totality, some new state of affairs may emerge in which both the world as we know it and our ideas about it may undergo an unending processes of yet further change.

With this we have in essence carried the presentation of our cosmology and our general notions concerning the nature of the totality to a natural (though of course only a temporary) stopping point. From here on we can further survey it as a whole and perhaps fill in some of the details that have been left out in this necessarily sketchy treatment before going on to new developments of the kinds indicated above.

Notes

1 Fragmentation and wholeness

1 See, for example, J. Krishnamurti, *Freedom from the Known*, Gollancz, London, 1969.

2 The rheomode – an experiment with language and thought

1 Actually, the Latin root 'videre' in 'divide' does not mean 'to see' but 'to set apart'. This appears to have come about in a coincidental manner. However, the purposes of the rheomode are served much better by taking advantage of this coincidence, and to regard division as primarily an act of perception rather than a physical act of separation.

2 Whenever a word is obtained from a form with a prefix, such as di-, co-, con-, etc., in the root verb of the rheomode, this prefix will be separated from the main verb by a hyphen, in order to indicate how the verb has been constructed in this way.

3 Note that, from now on, in the interests of brevity we will generally not give as full a description of the meaning of the root form as we have been doing thus far.

3 Reality and knowledge considered as process

1 A.N. Whitehead, *Process and Reality*, Macmillan, New York, 1933.

2 H. C. Wyld, *The Universal Dictionary of the English Language*, Routledge & Kegan Paul, London, 1960.

3 J. Piaget, *The Origin of Intelligence in the Child*, Routledge & Kegan Paul, London, 1953.

4 Hidden variables in the quantum theory

1 D. Bohm, *Causality and Chance in Modern Physics*, Routledge & Kegan Paul, London, 1957.
2 See J. von Neumann, *Mathematical Foundations of the Quantum Theory*, Princeton University Press, 1955; W. Heisenberg, *The Physical Principles of the Quantum Theory*, University of Chicago Press, 1930; P. Dirac, *The Principles of Quantum Mechanics*, Oxford University Press, 1947; P. A., Schilp (ed.), *Albert Einstein, Philosopher Scientist*, Tudor Press, New York, 1957, especially ch. 7 for a discussion of Bohr's point of view.
3 *Ibid.*
4 von Neumann, *op. cit.*
5 A. Einstein, N. Rosen, and B. Podolsky, *Phys. Rev.*, vol. 47, 1935, p. 777.
6 D. Bohm, *Quantum Theory*, Prentice-Hall, New York, 1951.
7 For a discussion of Bohr's point of view see Schilp, *op. cit.*, ch. 7.
8 D. Bohm, *Phys. Rev.*, vol. 85, 1952, pp. 166, 180.
9 L. de Broglie, *Compt. rend.*, vol. 183, 1926, p. 447 and vol. 185, 1927, p. 380; *Revolution in Modern Physics*, Routledge & Kegan Paul, London, 1954.
10 D. Bohm and J. V. Vigier, *Phys. Rev.*, vol. 96, 1954, p. 208.
11 For a more detailed discussion see Bohm, *Causality and Chance in Modern Physics*, ch. 4.
12 Bohm and Vigier, *op. cit.*; Bohm, *Causality and Chance in Modern Physics*.
13 Bohm, *Phys. Rev.*, vol. 85, 1952, pp. 166, 180; Bohm and Vigier, *op. cit.*; Bohm, *Causality and Chance in Modern Physics*.
14 Bohm and Vigier, *op. cit.*
15 Bohm, *Phys. Rev.*, vol. 85, 1952, pp. 166, 180; Bohm and Vigier, *op. cit.*; Bohm, *Causality and Chance in Modern Physics*.
16 G. Kallen, *Physica*, vol. 19, 1953, p. 850; *Kgl Danske Videnskab. Selskab, Matfys. Medd.*, vol. 27, no. 12, 1953; *Nuovo Cimento*, vol. 12, 1954, p. 217; A. S. Wightman, *Phys. Rev.*, vol. 98, 1955, p. 812; L. van Hove, *Physica*, vol. 18, 1952, p. 145.
17 *Ibid.*
18 Private communications.
19 Private communications.
20 Van Hove, *op. cit.*; private communications.
21 A similar result is obtained when one treats the large-scale properties of an aggregate containing a great number of interacting particles. One obtains collective properties (e.g., oscillations) that determine themselves almost independently of the details of individual particle motions. See D. Bohm and D. Pines, *Phys. Rev.*, vol. 85, 1953, p. 338 and vol. 92, 1953, p. 609.
22 This analogy was first shown by Fürth for the case of Brownian motion of a particle. See Bohm, *Causality and Chance in Modern Physics*,

ch. 4.
23 Bohm and Pines, *op. cit.*
24 M. Born, *Mechanics of the Atom*, Bell, London, 1927; H. Goldstein, *Classical Mechanics*, Addison-Wesley, Cambridge, Mass., 1953.
25 *Ibid.*
26 Born, *op. cit.*
27 Private communication.
28 For example, a synchronous electric motor tends to run in phase with the alternating current coming from the generator. There are innumerable other such examples in the theory of non-linear oscillations. A fuller discussion of non-linear oscillations is given by H. Jehle and J. Cahn, *Am. J. Phys.*, vol. 21, 1953, p. 526.
29 Born, *op. cit.*
30 Somewhat more general linear combinations can be taken but they only serve to complicate the expressions without changing the basic features of the problem.
31 D. Bohm and Y. Aharonov, *Phys. Rev.*, vol. 108, 1957, p. 1070.

5 *Quantum theory as an indication of a new order in physics. Part A*

1 This notion of order was first suggested to me in a private communication by a well-known artist, C. Biederman. For a presentation of his views see C. Biederman, *Art as the Evolution of Visual Knowledge*, Red Wing, Minnesota, 1948.
2 M. Born and N. Wiener, *J. Math. Phys.*, vol. 5, 1926, pp. 84–98; N. Wiener and A. Siegel, *Phys. Rev.*, vol. 91, 1953, p. 1551.
3 This notion has been discussed in chapters 1 and 3 from another point of view.
4 For a discussion of this point see D. Bohm, *Quantum Theory*, Prentice Hall, New York, 1951.
5 For an extensive discussion of this effect see *ibid*, ch. 22; for a later point of view on this subject see J. S. Bell, *Rev. Mod. Phys.*, vol. 38, 1966, p. 447.
6 N. Bohr, *Atomic Theory and the Description of Nature*, Cambridge University Press, 1934.
7 J. von Neumann, *Mathematical Foundations of Quantum Mechanics*, Princeton University Press, 1955.

6 *Quantum theory as an indication of a new order in physics. Part B*

1 For a very clear presentation of this view see T. Kuhn, *The Nature of Scientific Revolutions*, University of Chicago Press, 1955.
2 J. Piaget, *The Origin of Intelligence in the Child*, Routledge & Kegan Paul, London, 1956.
3 See D. Bohm, B. Hiley and A. Stuart, *Progr. Theoret. Phys.*, vol. 3, 1970, p. 171, where this description of a perceived content considered

as the intersection of two orders is treated in a different context.

4 See, for example, D. F. Littlewood, *The Skeleton Key of Mathematics*, Hutchinson, London, 1960.

5 See, for example, *ibid*.

7 *The enfolding-unfolding universe and consciousness*

1 See *Re-Vision*, vol. 3, no. 4, 1978, for a treatment of this subject in a different way. (Published at 20 Longfellow Road, Cambridge, Mass. 02148, USA.)

2 See D. Bohm, *Causality and Chance in Modern Physics*, Routledge & Kegan Paul, London, 1957, ch. 2, for a further discussion of this point.

3 For a more detailed discussion of this point see, for example, D. Bohm and B. Hiley *Foundations of Physics*, vol. 5, 1975, p. 93.

4 For a detailed discussion of this experiment see D. Bohm, *Quantum Theory*, Prentice-Hall, New York, 1951, ch. 22.

5 See D. Bohm, *Causality and Chance in Modern Physics*, ch. 2, for a discussion of this feature of 'indeterministic mechanism'.

6 See D. Bohm and B. Hiley *Foundations of Physics*, vol. 5, 1975, p. 93, and D. Bohm, *Quantum Theory*, Prentice Hall, New York, 1951, for a more detailed treatment of this feature of the quantum theory.

7 Mathematically one derives all the properties of the system from a 3*N*-dimensional 'wave function' (where *N* is the number of particles) which cannot be represented in three-dimensional space alone. Physically one actually finds the non-local, non-causal relationship of distant elements described above, which corresponds very well with what is implied by the mathematical equations.

8 Notably those in which the 'wave function' of the combined system can be factored approximately into two separate three-dimensional wave functions (as shown in Bohm and Hiley, *op. cit.*).

9 This is just an example of the combination of wavelike and particle-like properties of matter described in section 2.

10 This sort of calculation is suggested in D. Bohm, *Causality and Chance in Modern Physics*, Routledge & Kegan Paul, London, 1957, p. 163.

11 In section 8 we shall see that time, as well as space, may be enfolded in this way.

12 Compare with the idea of sub-system, system, and super-system, suggested in Bohm and Hiley, *op. cit.*

13 This notion has already been suggested in a preliminary way in chapter 3.

14 See Karl Pribram, *Languages of the Brain*, G. Globus *et al.* (eds), 1971; *Consciousness and the Brain*, Plenum, New York, 1976.

15 E.g., as shown in section 3, a linearly ordered array of droplets may be enfolded together in such a way that this order is still subtly held

in the whole set of ensembles of ink particles.

16 As shown in the appendix to chapter 6, on the implicate order the basic algorithm is an *algebra* rather than the calculus.

17 For a more detailed discussion, see D. Bohm, *The Special Theory of Relativity*, Benjamin, New York, 1965, Appendix.

18 See *ibid*.

19 This illusion is essentially the one discussed in chapters 1 and 2, in which the whole of existence is seen as constituted of basically static fragments.

20 In some ways this idea of an 'unconscious' background is similar to that of Freud. However, in Freud's point of view the unconscious has a fairly definite and limited kind of content and is thus not comparable to the immensity of the background that we are proposing. Perhaps Freud's 'oceanic feeling' would be somewhat closer to the latter than would be his notion of the unconscious.

21 This corresponds to the quantum theoretical requirement that electrons may go from one state in space to another, without passing through intermediate states.

Index